The Reacting Atmosphere 1

W0225694

Editor-in-Chief

Ralf Koppmann, Wuppertal, Germany

Series editors

Manfred Fischedick, Wuppertal, Germany
Michael Günther, Wuppertal, Germany
Martin Riese, Jülich, Germany
Peter Wiesen, Wuppertal, Germany

For further volumes:
http://www.springer.com/series/13396

The series *The Reacting Atmosphere* will present the objectives and visions of a new research network combining different disciplines involved in climate research. The objective of the network is to understand the highly complex regulatory cycles in the atmosphere taking into account all important parameters, to identify important atmospheric processes, to examine policies with respect to their consequences and, based on this to derive recommendations on how in a changing world targeted suggestions for improvement can be realized. The participating academic institutions will exploit synergies through joint research activities and make an important contribution to international research efforts directed at understanding climate change. The reader will learn about the activities and probably find a point of contact for future collaborations.

Ralf Koppmann

Editor

Atmospheric Research From Different Perspectives

Bridging the Gap Between Natural
and Social Sciences

Editor
Ralf Koppmann
Faculty of Mathematics and Natural
  Sciences
Department of Physics
University of Wuppertal
Wuppertal
Germany

ISSN 2199-1138   ISSN 2199-1146 (electronic)
ISBN 978-3-319-06494-9   ISBN 978-3-319-06495-6 (eBook)
DOI 10.1007/978-3-319-06495-6
Springer Cham Heidelberg New York Dordrecht London

Library of Congress Control Number: 2014938718

Mathematical Subject Classification (2010): 37-XX, 62-XX, 65-XX, 70-XX, 76-XX, 86-XX, 91-XX, 92-XX, 93-XX

Printed on acid-free paper

Springer is part of Springer Science+Business Media (www.springer.com)

Preface

For millions of years our atmosphere has been in a state of constant change, and it continues to change today. Most of the changes are triggered by natural processes and not influenced by human beings. But some significant changes primarily observed during the last century, such as the increasing levels of greenhouse gas emissions, are obviously due to human activities. Since there is no other habitat for us or future generations but this planet, it is our responsibility to understand these changes and investigate their causes. Only if reliable predictions about the future development of our atmosphere are possible, can we develop reasonable solutions without creating new and even more serious problems elsewhere.

In a changing world our traditional approaches to solving environmental problems, taken alone, are doomed to fail. Accordingly, new perspectives and new strategies are needed to prevent us from destroying ourselves. Anthropogenic climate change and the interaction with air quality endanger the social and economic basis of all people around the world. This implies an urgent need to promptly develop efficient, sustainable mitigation and adaptation strategies in response to the damage already done.

As the editors of the highly recommended book "Interdisciplinarity and Climate Change" (see Further Reading, Chap. 1) point out, it is necessary to create a "framework for coherently integrating the findings of distinct sciences, on the one hand, and for integrating those findings with political discourse and action, on the other" ... so as to find "... ways to conceptualise and measure relationships between social activities and climate outcomes in pursuit of reduction in greenhouse gases."

This first volume of the series "The Reacting Atmosphere" is a point of departure. It presents the idea of approaching air quality and climate change from a "systemic view," describes the objectives and strategy of the Research Network and provides a brief overview of the developments over the last several years in the context of establishing the Research Network. In the volumes to come we will report on the results of current projects and the progress of the Network. By pursuing this approach, we hope to attract researchers from the various disciplines to join our efforts, and to spark the interest of those researchers with expertise not yet included in our Network. We further aim to foster transdisciplinary research and initiate new projects to improve our systemic view. Accordingly, contributions from researchers working in these areas are very welcome.

We are aware of the fact that this project is extremely challenging and that our goals are very ambitious. However, we feel that it is high time to turn the idea into a reality by combining competences in atmospheric physics and chemistry, applied mathematics, and socio-economic science, allowing us to go beyond the current state of the art in understanding the role of the atmosphere in global change. We believe that it is time for us as scientists at the forefront of atmospheric research to more openly and clearly communicate our findings to political decision makers and the general public, and to assess actions and measures taken. And we feel that it is time to disseminate our knowledge to all areas of education. Join us for the journey of achieving these goals.

Wuppertal, March 2014   Ralf Koppmann

Contents

Chapter 1
"The Reacting Atmosphere": A Systemic Approach to Atmospheric Research

Ralf Koppmann

The Research Network

The Research Network "The Reacting Atmosphere" is a trans-disciplinary consortium of natural and social scientists, economists and mathematicians. Its objective is to acquire a systemic view of the role of the atmosphere for air quality and climate change. This systemic view is mandatory to understand the interactions of atmospheric change with political measures, societal and economic developments and their feedbacks.

Motivation

In a changing world our traditional approaches for solving environmental problems, taken alone, are doomed to fail. Accordingly, new thinking and new actions are needed to prevent us from destroying ourselves. Human-made changes of climate and the interaction with air-quality endanger the social and economic basis of all people around the world. This implies a strong need to promptly develop efficient sustainable mitigation and adaption strategies in response to the already caused changes.

The Intergovernmental Panel on Climate Change (IPCC) has identified the interactions between atmospheric composition (e.g. air quality) and climate to be a key uncertainty in our understanding of climate change, in particular on decadal time scales. Therefore, it is of utmost importance to gain a sound understanding of the complex global and regional interdependencies between human activities, climate and air-quality. This requires the development of a systemic view that

R. Koppmann (✉)
Faculty of Mathematics and Natural Sciences, Physics Department,
University of Wuppertal, Wuppertal, Germany
e-mail: koppmann@uni-wuppertal.de

R. Koppmann (ed.), *Atmospheric Research From Different Perspectives*,
The Reacting Atmosphere 1, DOI: 10.1007/978-3-319-06495-6_1,
© Springer International Publishing Switzerland 2014

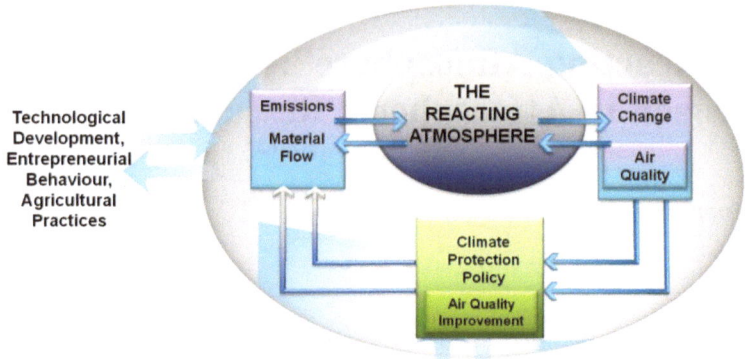

Fig. 1.1 Systemic view of air quality and climate change including the interactions of atmospheric change with political measures, societal and economic developments and their feedbacks

includes societal and economic processes as well as physical and chemical processes in the atmosphere (cf. Fig. 1.1), leading also to improved predictive capabilities and normative implications.

Our Vision: The Systemic View

Today, the scientific community agrees that global temperatures will increase in the next 100 or 200 years, although the predictions based on the results of various climate models under the consideration of different scenarios show tremendous variations, reaching from an increase of the mean global temperature of 2–6.5 °C until the end of this century. Despite of this general agreement nobody knows how the Earth system will respond to these changes. We are not even rudimentary able to obtain a conclusive statement, what the feedbacks on land use, agricultural practices, and energy or food supply will be.

Our research aims at a quantitative understanding of the key processes of the system as a whole, both the interdependencies of atmospheric processes and parameters and the interaction between physico-chemical and socio-economic processes. This quantitative understanding will help to assess previous political and technical measures and will provide the basis for the development of new sustainable strategies for economic and societal developments.

Only a systemic approach can initiate technology and policy measures as well as societal and educational developments to improve air quality and counteract climate change in a sustainable manner, in Europe and worldwide.

Our vision is to achieve this objective by a trans-disciplinary research approach covering atmospheric sciences, scientific computing and mathematics as well as economic and social sciences.

Why a "Systemic View"?

The Earth system as a whole is what we would call a really complex system. General attributes of complex systems are the large numbers of characteristics, which are not easily accessible, which develop individually in a very dynamical way and which mutually interact. Another property of complex systems is a—sometimes large—time delay between cause and effect. Generally, from their evolutionary development human beings are not suited to handle complex systems. Humans usually tend to think and act in a linear way and, thus, to solve problems systematically; one step after the other, assuming that one problem has only one solution. In doing so, immediate effects may be considered. Cascading effects, feedbacks or different specific effects as a result of a singular cause are usually not considered at all. Our causality understanding, which is obviously an inherited behaviour, presupposes some principal rules: (1) the cause occurs before the effect, (2) cause and effect occur together, (3) cause and effect occur close in time and space and (4) cause and effect resemble each other. This human pattern of behaviour makes it difficult if not impossible to act well-directed in situations where at a first glance causes and effects seem to be not directly related to each other or the time lag between causes and effects is counted in decades or centuries.

Another aspect to solve problems in complex systems is that usually the seemingly most important problem is solved first, very often controlled by its conspicuity or by the available competences. If problems encompass different disciplines, often the methods of one subject area are used to solve problems that are outside one's subject area. And, last but not least, the importance to monitor the results and the reactions of a system to a specific measure is often neglected.

These are exactly the main problems with processes such as climate change. The temporal and spatial course of processes within the Earth system following global change is not within the reach of the rules stated above. In addition, there are many disciplines involved, and monitoring effects need a lot of time, large efforts and sometimes cause high costs. Although numerical models as well as their spatial resolution were getting significantly better in the last years and computing power increased considerably, the predictability of future developments regarding global change is still extremely low and uncertainties are high. Nevertheless, assessing developments in the past should make us aware of the fact that we face complex problems. We have no choice but to accept the challenge and to learn how to deal with this complex system and its compulsory complex solutions.

The Case of Chlorofluorocarbons

The invention of chlorofluorocarbons (CFCs) in the 1920s may be taken as an example. These chemicals have excellent properties: they are nontoxic, chemically inert, inflammable, and offer a wide field of applications such as refrigerants, foaming agents and degreasing solvents. 50 years after the start of industrial production and use of these compounds James Lovelock, a British chemist, invented the electron capture detector, a special instrument to detect halogenated trace gases in the atmosphere. While measuring the CFCs in the atmosphere he quickly recognized that the total atmospheric burden corresponded more or less to the total amount produced up to that time. A few years later Sherwood Rowland and Mario Molina, two atmospheric chemists at the University of California, detected that the only way to remove these compounds from the atmosphere is the transport from the troposphere into the stratosphere followed by the breakdown of the molecules due to the solar ultraviolet radiation. This process, however, led to the production of chlorine, which in turn in a catalytic process is able to destroy ozone. Rowland and Molina warned that the increasing concentrations of CFCs will lead to a significant destruction of the ozone layer, which protects all life forms (including humans) against ultraviolet radiation. However, this warning was not taken seriously. Another 10 years later the ozone hole over Antarctica was discovered. Following four years of intensive research, halogens such as chlorine and bromine originating from CFCs were identified to be the cause of ozone destruction. These findings finally led to the Montreal Protocol, which entered into force in January 1989, banning the production and use of these environmentally hazardous compounds.

On the one hand, this is a classic example for the temporal and spatial behaviour of causes and effects in atmospheric processes. The time lag between the invention of CFCs and the effect in form of severe ozone depletion is counted in decades. While the vast amount of these compounds were emitted in the northern hemisphere, the most significant and visible effect occurred over Antarctica. On the other hand, this is an optimistic-provoking example for the capability of mankind to act well-directed under the urge to save the protective shield against the life-destructive ultraviolet radiation.

The Time Delay of Climate Effects

Climate effects behave in a very similar way. We are significantly increasing the concentration of greenhouse gases. With the change in land use we change the albedo of the Earth's surface. With increasing emissions of other gases and particulate matter we change the radiative transfer in the Earth's atmosphere. As a consequence, the temperature started to increase and will continue to do so. It may take decades before noticeable and hazardous effects will occur. Regarding the oceans, reaction times are even longer, in some instances hundreds of years. Some of these reactions will have positive feedbacks still increasing the impacts. And some of these reactions will lead to so called tipping points, where parts of our Earth system will irreversible migrate into a new steady state. The loss of the rain forests due to continuous droughts, the deglaciation of the northern hemispheric ice cap, or the disappearance of the Asian monsoon would be such tipping points.

Climate Reductionism

There are a number of key issues in the context of the complex phenomena connected with climate change and our response to it. Presently, there are a number of weaknesses in our behaviour regarding the obvious changes in the Earth system.

Greenhouse gases are naturally and anthropogenically emitted infrared active trace gases in the atmosphere. The most important greenhouse gas is water vapour, the concentration of which is affected by climate change leading to a so called positive feedback mechanism. A warmer atmosphere can take up more water vapour leading to an increase in global warming. Other greenhouse gases are carbon dioxide (CO_2), methane (CH_4), nitrous oxide (N_2O) and ozone (O_3). Additionally, there is a large variety of entirely anthropogenic greenhouse gases like SF_6, the chlorofluorocarbons (CFC), and a number of other gases. The global warming potentials of these compounds are by orders of magnitude larger than those of the also naturally emitted gases. The fully halogenated compounds were abandoned by the Montreal Protocol and the follow up protocols leading to significant depletions of the gases since then. CO_2 and a few other greenhouse gases are under the restrictions of the Kyoto Protocol with more or less questionable results.

However, despite of these facts and the meanwhile profound knowledge of the concentrations and distribution of these gases and their effects on the Earth system, the discussion of necessary political and societal responses to climate change with respect to environmental issues, changes in energy "production" and consumption, transport strategies, is usually reduced to the carbon dioxide emissions.

Integrating Natural and Social Scientific Work

If we talk about a "complex system" such as the Earth system we actually have to be aware that we have to deal with two different types of complexity, a structural complexity and a behavioural complexity. The structural complexity of a system increases with an increasing number of components and interactions. This type of complexity can be described by quantitative characteristics. The number of elements or subsystems and the number of interconnections are measurable and can eventually be described by a reduction to key parameters determining the interactions.

Behavioural complexity has a qualitative characteristic. Here properties of the system as a whole and the relationships and processes between the subsystems have to be considered. These relationships and processes are characterised by a non-linear behaviour. They are dependent of their previous history, and, usually, not predictable. Furthermore, the structure and the state of the system can be significantly altered by dynamical interactions as well as feedback mechanisms. This becomes especially noticeable as soon as human interactions come into play. These interactions cannot be described by a linear approach and simple "cause and effect" processes, because they occur on different temporal and spatial scales. Furthermore, local interaction can lead to emergent phenomena in the global structure of the system as a whole.

To account for these facts, both quantitative and quantitative approaches to an understanding of the system as a whole have to be considered. This means, natural sciences describing physical and chemical processes and interactions are necessary but not sufficient to increase our understanding of the system. Social and economic science has to be included in order to investigate and assess also the interactions between humans and the environment.

All present studies and projects mainly focus on natural scientific work, investigating the on-going changes in our atmosphere and the processes responsible for these changes. If ever, an integration of social or economic science is only realized to a very small extent. The transfer of theoretical knowledge into practical actions has not at all been investigated up to now. The question, if researchers give the right information in the right way to the right addressees is not even partially answered. At the same time an assessment, if and how decision makers make use of scientific results in the development of actions and measures is also lacking, as is an assessment of the measures taken so far. The question in which way specific measures affect the atmosphere and, even more important, if these measures probably produce unwanted trade-offs, seems not to be asked.

To understand today's climate it is, of course, necessary to understand the development of climate in the past. However, this knowledge is not simply transferable to the future. Small variations in the initial conditions, incalculable positive or negative feedbacks, unforeseeable variations in external forcings, and unpredictable human behaviour may considerably change the development of the system. Any decision or measure aiming at a (positive) influence on the climate system depends on a vast variety of processes, parameters and interactions. In a complex and non-linear system it is impossible to generate planning reliability. However, predictions that usually are based on different potential scenarios can be improved if we take into account dynamical interactions as well as non-linear behaviour, and also simulate possible surprising effects. In addition, it is indispensable to continuously assess measures and decisions regarding their impact on air quality and climate.

One of the objectives of the Research Network is the development of concepts to successfully build a bridge between natural and socio-economic sciences.

The Systemic Approach

Our research aims at a as far as possible quantitative understanding of the key processes of the system as a whole, both the interdependencies of atmospheric processes and parameters and the interaction between physico-chemical and socio-economic processes. This will help to assess previous political and technical measures and will provide the basis for the development of new sustainable strategies for economic and societal developments. Only a systemic approach can initiate technology and policy measures as well as societal and educational developments to improve air quality and counteract climate change in a sustainable manner, in Europe and worldwide.

Our vision is to achieve this objective by a trans-disciplinary research approach covering atmospheric sciences, scientific computing and mathematics as well as economic and social sciences. Thus, we need to apply a highly non-linear, cross-linked thinking to take the challenge of counteracting climate change. We need a new degree of abstraction. We have to abandon well established strategies and break new grounds.

An Example of Unconsidered Interdependencies

The conflict between air pollution mitigation measures, economic and technological measures and chemical processes in the atmosphere is shown taking the air pollutant nitrogen dioxide (NO_2) as an example. The NO_2 issue became evident through the introduction of new European limiting values in January 2010, which are currently still exceeded in many European areas. The reason for this is manifold and partly caused by unconsidered interdependencies:

- The introduction of continuously regenerating particle traps, e.g. in public busses in the city of London, led to reduced particle emissions but unexpectedly to increased emissions of nitrogen dioxide (NO_2).
- During the economic crisis in 2008 Germany subsidised purchasing new cars in order to support car manufactures. Accordingly, the introduction of low emission vehicles, such as EURO-6 cars, into the car fleet is now delayed several years. Thus the compliance of new NO_2 limiting values will be achieved much later than expected.
- Not even taken into account is the fact that increasing NO_2 concentrations are also caused by the yet unexplained increase of background ozone concentrations.

The example of NO_2 shows impressively in a nutshell the impact of so far disregarded interdependencies on the "reacting atmosphere" and the necessity of a systemic approach.

A First Step Towards a Systemic Approach

Currently, hydrogen (H_2) is intensively discussed as a major element of future energy supply in order to reduce the emission of greenhouse gases such as carbon dioxide (CO_2) and methane (CH_4). However, the production and supply chain of hydrogen in a future global hydrogen economy is inevitably accompanied by leakages leading to an increase of hydrogen in the atmosphere. As a consequence, this could yield a significant increase of water vapour concentrations in the stratosphere, which impacts the ozone layer and surface climate, two areas of high societal relevance.

Previous studies focused only on particular aspects, but did not consider the whole supply chain and its impacts. Therefore, the possible impact of a future hydrogen economy was investigated as a first attempt in a systemic approach. A sophisticated atmospheric model was combined with available scenarios for future economic, population growth and implementation of a future hydrogen economy. In addition, the range of possible production and leakage rates of hydrogen was analysed in a technical assessment considering the complete process chain from production to utilization.

In contrast to previous studies the systemic approach showed that the impact of a future hydrogen economy on stratospheric ozone will be most likely of minor importance. Thus, hydrogen may be used as a promising future energy carrier. This example shows that only through the systemic approach a more realistic assessment of the potential impacts of a hydrogen economy was achieved.

The Strategy of the Research Network

More than half of mankind is living in urban areas being responsible for almost 80 % of global greenhouse gas (GHG) emissions. To cut emissions, technological solutions are not sufficient as their effects may be overcompensated by economic growth and rebound effects. Therefore achieving ambitious GHG reduction goals requires a holistic approach, including the consideration of consumer behaviour and social innovations as important topics. Additionally, urban climate protection strategies can only be successful if they pay attention to additional ecological (e.g. air quality) and societal targets besides the climate issue.

Being aware that a systemic view is not an easy task, but inevitable to enhance our understanding and prediction capabilities, the Research Network will start by focusing its research activities on two regions.

The Rhine-Ruhr Metropolitan Area as a First Focus

One research focus will be the Rhine-Ruhr Metropolitan Area as a typical example of a large sprawling urban area. This region, which is possibly the biggest European agglomeration, is comparably well understood in terms of anthropogenic emissions and looks back on a 50 years record of corresponding political and technological measures. In addition, the Research Network "The Reacting Atmosphere" can build on a high-quality database for air quality issues as well as a profound understanding of the socio-economic system in this area. It is thus an ideal starting point for the development of a systemic approach.

Innovation City

Shaping sustainable urban infrastructures is a complex process and experience is scarce. The project "Innovation City Ruhr" was started as a multi-dimensional and transdisciplinary real world experiment, which has a crucial role in the transition cycle. The aim is to learn more about achievable urban transition goals, socio-economic system interactions, alternative transition options and their characteristics and impacts.

In November 2010 the city of Bottrop located in the middle of the industrial heart of Germany was appointed "Innovation City Ruhr". The aim is to cut GHG emissions in a representative district with about 69.000 inhabitants by 50 % within 10 years. Additionally "better living" conditions shall be achieved.

Compared to many other projects "Innovation City" is going a step further. Society and technology are inextricably intertwined and science has the role to support the transition process, to help developing experiments and to trigger learning projects. The comprehensive accompanying research programme is being organized by one partner of the Research Network. The project brings together different disciplines and experiences. It works as a transmission belt between researchers, the planning team, investors, and political decision makers. Thus, research provides first-hand information and insights, which are absolutely crucial for national and international transfer of practical experience.

East Asian Megacities: Towards a Quantitative Understanding of Chemistry-Climate Interactions

The results from the Rhine-Ruhr Metropolitan Area will be used for corresponding investigations in the East Asian megacities. There, population growth and economic developments lead to severe air pollution being significantly higher than in any European area. In addition, this region has a significant impact on global air quality-climate interactions. This is based on emissions of greenhouse gases and their precursors combined with a special meteorological condition known as the Asian monsoon circulation, which amplifies the impact of these emissions.

Recent studies show that this circulation leads to rapid transport of large amounts of pollutants from the lower to the upper atmosphere (up to 20 km), which is extremely climate sensitive. From there, pollutants are readily dispersed around the globe.

A good example for the importance of the Asian monsoon circulation is its impact on stratospheric aerosols (sulphuric acid droplets). This aerosol was recently observed to increase, though the global sulphur emissions are decreasing. The reason is obviously the tremendous increase of sulphur dioxide (SO_2) emissions in densely populated regions of China since 2002, most likely in connection with the effective transport into the stratosphere via the Asian Monsoon circulation.

This convincing example underlines the importance of investigating physico-chemical processes in the atmosphere simultaneously with socio-economic processes determining important emissions of greenhouse gases and their precursors.

Further Reading

1. Archer, D.: Global Warming—Understanding the Forecast. Wiley-Blackwell, Oxford (2012)
2. Archer, D., Pierrehumbert, R. (eds.): The Warming Papers—The Scientific Foundation for the Climate Forecast. Wiley-Blackwell, Oxford (2011)
3. Bhaskar, R., Frank, Ch., Hoyer, K.G., Naess, P., Parker, J. (eds.): Interdisciplinarity and Climate Change—Transforming Knowledge and Practice for Our Global Future. Routledge, London (2010)
4. Edwards, A.R.: The Sustainability Revolution—Portrait of a Paradigm Shift. New Society Publishers, Gabriola Island, Canada (2008)
5. Feck, Th.: Wasserstoff-Emissionen und ihre Auswirkungen auf den arktischen Ozonverlust—Risikoanalyse einer globalen Wasserstoffwirtschaft, Schriften des Froschugnszentrums Jülich. Energie und Umwelt, Band 51 (2009)
6. Ison, R.: Systems Practice: How to Act in a Climate-Change World. Springer, London (2010)
7. Leggewie, C., Welzer, H.: Das Ende der Welt, wie wir sie kannten. Fischer-Verlag, Frankfurt (2009)
8. McKibben, B.: Eaarth—Making a Life on a Tough New Planet. Times Books, New York (2010)
9. Ruddiman, W.F.: Earth's Climate—Past and Future. W.H. Freeman, New York (2008)
10. Ruddiman, W.F.: Plows, Plagues and Petroleum–How Humans Took Control of Climate. Princeton University Press, Princeton (2005)
11. Schmidt, G., Wolfe, J.: Climate Change—Picturing the Science. W.W. Norton, New York (2009)
12. Weart, S.R.: The Discovery of Global Warming. Harvard University Press, Cambridge (2008)

Chapter 2
Interdependencies of Atmospheric Processes

Peter Wiesen

One task of the research network is to combine the key competences of the partners, namely laboratory experiments and field investigations are combined. In particular, this aims to answer the following key research questions, which are of paramount importance to significantly improve the understanding of the composition change of the Earth's atmosphere and its impacts:

- How does the dispersion and conversion of atmospheric pollutants occur in different atmospheric environments?
- What are the key processes responsible for the removal of atmospheric pollutants, both gases and particles, in different atmospheric environments?
- How does the exchange of air masses work between different atmospheric layers, namely troposphere and stratosphere on regional and global scales?

One key parameter for a full understanding of atmospheric processes is the removal of atmospheric pollutants, e.g. the atmospheric oxidation capacity. Knowing the key compounds and processes determining the removal of pollutants is the basis for advanced policy measures to improve air quality and to reduce the impact of atmospheric pollutants on climate in a timely manner. Another key process is the exchange of air masses between different atmospheric layers. Only a detailed understanding of the coupling of local and regional emissions of pollutants with their distribution on a larger scale, i.e. the mid and upper troposphere, but also the lower stratosphere, will allow the development of strategies to minimise the impact of local emissions (e.g. from megacities) on climate.

The vision is to determine all key parameters influencing atmospheric processes on different temporal and spatial scales and the impact on air quality and climate, taking into account both anthropogenic and biogenic emissions. This will enable physical and chemical processes in the atmosphere to be described qualitatively

P. Wiesen (✉)
Faculty of Mathematics and Natural Sciences, Chemistry Department,
University of Wuppertal, Wuppertal, Germany
e-mail: wiesen@uni-wuppertal.de

R. Koppmann (ed.), *Atmospheric Research From Different Perspectives*,
The Reacting Atmosphere 1, DOI: 10.1007/978-3-319-06495-6_2,
© Springer International Publishing Switzerland 2014

and quantitatively. We particularly strive for a detailed understanding of the interdependencies of key atmospheric processes and parameters and their direct impact on air quality and climate.

The mid-term objectives are the determination of detailed source apportionments and a profound understanding of the underlying physico-chemical processes leading to specific distributions of atmospheric trace compounds. The results will be used to analyse the high complexity of the interdependencies of the various subsystems in order to develop new measurement strategies and control systems in a purposeful and target-oriented manner.

The long-term objectives are the quantification of the turnover of atmospheric pollutants, their transport from the planetary boundary layer to the upper troposphere/lower stratosphere and a profound understanding of the formation of secondary pollutants as well as the deduction of improved parameters describing radiative forcing.

During recent years it has been conclusively demonstrated that regional air quality and global climate change are highly interrelated because emissions of many pollutants affect both air quality and climate change. Furthermore, the fundamental chemistry affecting air quality and global climate is somewhat similar.

In addition to the greenhouse gases such as carbon dioxide (CO_2), nitrous oxide (N_2O) and methane (CH_4) classical air pollutants take a significant influence on the climate. These are carbon monoxide (CO), nitrogen oxides (NO_x), ammonia (NH_3), sulphur dioxide (SO_2), volatile organic compounds (VOCs), particulate matter and soot/black carbon. They are all short-lived, but have a significant impact on the chemistry of the atmosphere, for example, to the formation of ozone (O_3) and aerosols. Ozone and aerosols are directly affecting the climate. These air pollutants, which have attracted relatively little attention in the climate debate, come from a variety of sources such as agriculture, transport, energy production including heating, but may also be of biogenic origin.

Recent studies show that a reduction of methane, ozone and soot in the atmosphere could slow the temperature rise in the coming decades. We may buy us time if we do everything we can to improve our air quality so that we emit less soot and methane and minimize the formation of ozone through a further reduction of VOCs and/or NO_x.

Fine particulate matter and nitrogen dioxide (NO_2) are the key problems for increasing air quality in Europe. Whereas particulate matter and the exceedance of PM limiting values have attracted considerable public attention during the last couple of years, the NO_2 problem is a relatively new one, which became mature through the introduction of new European limiting values in January 2010.

The reduction of nitrogen oxide ($NO_x = NO + NO_2$) emissions has been historically one of the key objectives for improving air quality in Europe. NO_x emissions have started to decrease considerably since the mid-eighties of the last century in many European areas.

In Germany, nitrogen oxide emissions are still largely caused by road traffic. Nationwide, the share of road transport in the total NO_x emissions is currently around 60 % [1]. Since the mid-80s, a significant decrease in NO_x emissions from road transport has been reported from Germany. According to emission calculations of the German Federal Environmental Agency, the decrease in total NO_x emissions between 1990 and 2005 is approximately 60 % [1]. Figure 2.1 shows an example of NO and NO_2 concentrations measured on major roads in the German Federal State of North Rhine-Westphalia (NRW) for the period 1990–2012 [2].

One can see a significant decline in the annual averages of NO_x pollution from about 80 to about 40 ppbv. This decrease is in agreement with the calculated NO_x emission trends. In contrast, the NO_2 concentrations in NRW stagnate in the same period at about 23 ppbv. This trend has been also observed nationwide [3] and in other European countries [4–7].

Nitrogen dioxide is already a problem for many cities due to its toxicity and key role in the formation of tropospheric ozone [8]. In Germany in 2010, still more than half of the major roads were well above the current annual limit of 40 $\mu g/m^3$ and approximately 20 ppbv, respectively [3]. Because of the problems of many EU member states in complying with the new NO_2 annual concentration limit, the European Commission introduced time extensions to meet limits until January 1, 2015. Since early 2004, the Physical Chemistry Laboratory of the Faculty for Mathematics and Natural Sciences of the University of Wuppertal, a partner in the research network, in cooperation with the State Office for Nature, Environment and Consumer Protection (LANUV) of the German federal state of North Rhine-Westphalia performed extended pollution measurements at two monitoring stations in Wuppertal and Hagen in order to clarify the reason for the almost stagnant NO_2 pollution [9].

Generally, in combustion processes, such as in engines of motor vehicles, NO_2 and NO are primarily formed and emitted directly. The directly emitted NO is converted in the atmosphere, partly by O_3 or peroxy (RO_2) radicals into NO_2, which is called in the following secondary NO_2:

$$NO + O_3; \; (RO_2) \rightarrow NO_2 + O_2; \; (RO) \tag{2.1}$$

In the presence of sunlight, NO_2 is partially photolysed back to NO:

$$NO_2 \xrightarrow{sunlight(O_2)} NO + O_3 \tag{2.2}$$

Assuming that the RO_2 photochemistry is negligible at the polluted urban kerbside stations investigated and that the background O_3 concentration is constant, the level of oxidants (OX) is given by [10]:

$$NO_2 + O_3 \equiv OX \tag{2.3}$$

The reason for the observed NO_2 trend is twofold. Firstly, the NO_2/NO_x *emission ratio* has increased significantly during the last two decades. Furthermore, caused by the nonlinear dependency on the NO_x level, *secondary NO_2* is decreasing much more slowly than expected from the decreasing NO_x levels. A detailed analysis of the data at two monitoring stations in Germany confirmed that

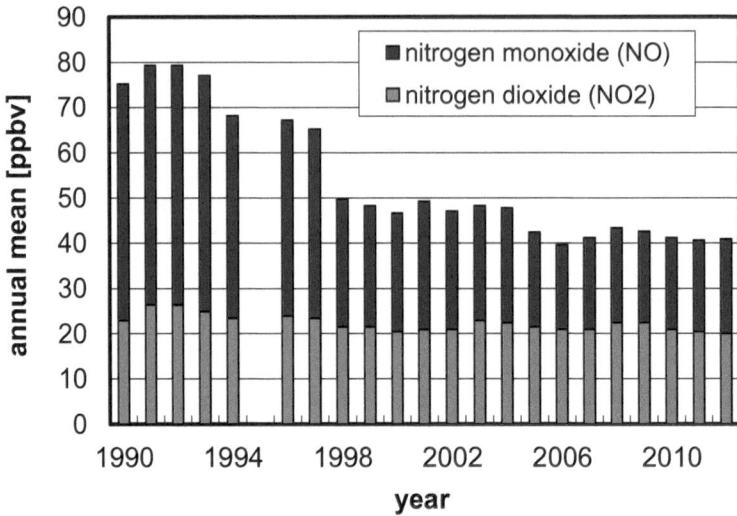

Fig. 2.1 NO_x and NO_2 concentrations measured on major roads in the German Federal State of North Rhine-Westphalia (NRW) for the period 1990–2012

the NO_2 concentrations are mostly determined by secondary NO_2 formation, through the reaction of NO with ozone [9].

In Fig. 2.2, simple box-model calculations are shown, in which the photostationary state NO_2 mixing ratio is calculated for a background ozone level of 40 ppbv and four different NO_2/NO_x emission ratios as a function of NO_x.

Only the NO_2 photolysis, reaction (2.2) and the ozone titration by NO, reaction (2.1), were considered. The rate coefficient for reaction (2.1) was taken from Atkinson et al. [11], and for the photolysis frequency of NO_2, a typical noontime summer value of $J(NO_2) = 8 \times 10^{-3} \text{ s}^{-1}$ was used. Since secondary NO_2 is also formed in the atmosphere in the presence of hydrocarbons and sunlight through RO_2 chemistry (see above) and since $J(NO_2)$ values (NO_2 sink reaction) are often lower than used here, the actual NO_2 concentrations may be even higher.

Figure 2.2 shows a highly nonlinear dependence of the steady state NO_2 level with NO_x, which is explained by the linear reaction kinetics of the NO_2 photolysis, sink reaction (2.2), and the second order kinetics of the NO_2 source reaction (2.1). An important conclusion from this result is that a reduction of the NO_x levels, e.g., by a factor of two will result in a much smaller reduction in the NO_2 mixing ratios, as long as the NO_x levels are significantly higher than the background O_3 levels. It is apparent that a further significant reduction of NO_x emissions is prerequisite to meet the current limit value for NO_2, almost regardless of the NO_2/NO_x emission ratio! For example, the NO_x annual average mixing ratio in Hagen for 2007 of 92 ppbv or for Wuppertal of 55 ppbv has to be reduced to ca. 35 ppbv. This result is also consistent with other studies from Germany suggesting a further reduction of NO_x emissions of at least 50 % in order to achieve the required reduction of the NO_2 concentration in the atmosphere [12, 13].

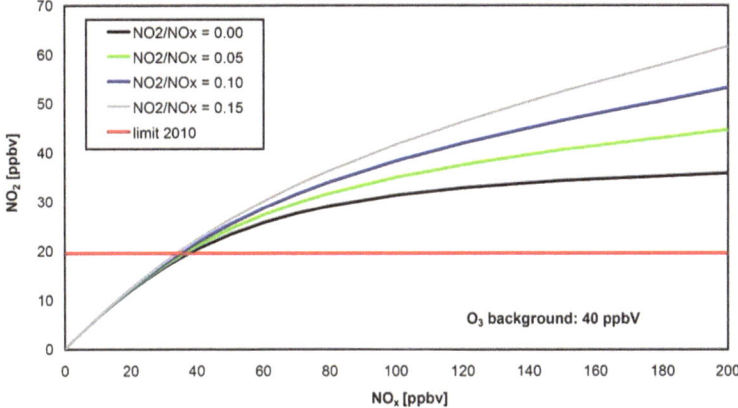

Fig. 2.2 Photostationary mixing ratio of NO_2 as a function of the NO_X mixing ratio, shown for different NO_2/NO_x emission ratios. Ozone background 40 ppbv, NO_2 photolysis rate 8×10^{-3} $[s^{-1}]$ representing typical conditions during summertime in Germany

A reduction of the primarily emitted NO_2 due to improved emission control systems alone is not sufficient to reduce the NO_2 concentrations significantly. Compliance with the NO_2 annual limit of approximately 20 ppbv requires a further drastic reduction of NO_x emissions in the near future. A similar conclusion has been drawn also for other European Countries [5, 7]. Thus, the exceedance of NO_2 limit values will remain a European problem within the next couple of years [14].

A relatively new technique for the reduction of NO_x in heavily polluted areas is currently quite controversially discussed, namely photocatalysis by e.g. using TiO_2-doped paints. Titanium dioxide (TiO_2) is a well-known photocatalyst, which is used to decompose pollutants both, in the gas and liquid phases [15–17]. TiO_2 exists in different modifications from which the anatase form is most active for photocatalytic reactions [18]. Presently only a limited number of studies is known, in which photocatalytic reactions of nitrogen oxides on TiO_2 were studied. An excellent overview can be found in the paper from Laufs et al. [19].

During the European PICADA project photocatalytic paints and mortar panels were studied in the laboratory and in a simulated street canyon [20, 21]. During the latter experiment high reduction of nitrogen oxides of 40–80 % was observed. However, these high numbers can be explained by the unrealistic high surface to volume (S/V) ratio of the simulated street canyon, which is a rate determining parameter in heterogeneous reactions.

In a very recent study by Langridge et al. [22], significant formation of HONO was observed in the photocatalytic reaction of NO_2 on self-cleaning window glass, from which the widespread use of photocatalytic materials could be questioned.

In the recent paper by Laufs et al. [19] photocatalytic *paint surfaces* have been shown to be an effective sink of NO, NO_2 and HONO under irradiation. The authors reported heterogeneous dark formation of HONO on non-catalytic paint surfaces and concluded that no additional negative impact by HONO formation on photocatalytic paints is expected.

Since NO, NO$_2$ and HONO are either directly harmful or indirectly lead to the formation of harmful secondary products, such as peroxy acetyl nitrate (PAN) or O$_3$, the use of photocatalytic *paint surfaces* has the potential to reduce the concentration of these harmful trace gases in the atmosphere.

As final end product of photocatalytic reactions of nitrogen oxides, the formation of adsorbed HNO$_3$/NO$_3^-$ with near to unity yield has been reported [19]. On urban photocatalytic active surfaces, for example on facades, HNO$_3$/NO$_3^-$ will be removed then by rain and could lead to acidification and eutrophication of the waste water, which may be reduced by adequate waste water treatment.

It is worth mentioning, that also in the atmosphere, nitrogen oxides are almost quantitatively oxidized into nitric acid mainly by the gas phase reaction of NO$_2$ with OH radicals during daytime and removed from the atmosphere by dry or wet deposition. In particular, dry deposition of gas phase HNO$_3$ has a negative impact on plants, or a direct impact on human health. Thus it may be speculated that the use of photocatalytic paints reduces also the negative impact of HNO$_3$/NO$_3^-$ on the environment.

The exact quantification of the reduction of nitrogen oxides in the urban environment by the use of photocatalytic paints is very difficult. Fast photochemical reactions as reported by Laufs et al. [19] were already nearly the diffusion limit in the experimental set-up used by the authors. In the real atmosphere, for example in a much larger street canyon, photochemical reactions will be also limited by transport (convection, diffusion). Accordingly, the exact quantification of the NO$_x$ reduction requires 3-D model calculations including the micrometeorology and the geometry of the corresponding street canyon.

However, Laufs et al. [19] reported a photocatalytic reduction of nitrogen oxides in a typical street canyon of only ca. 5 %, which was based on a simple estimation of the surface to volume ratio in a real street canyon in comparison with data from the PICADA experiment [20, 21].

Although a 5 % reduction seems to be quite small, this has to be compared with NO$_x$ reduction through other activities used to keep the new EU limit value for NO$_2$. For example, the city of Cologne, Germany reported an average NO$_2$ reduction of only 1.5 % after the implementation of a low emission zone [23].

References

1. Tappe, M., Friedrich, A., Höpfner, U., Knörr, W.: Berechnung der direkten Emissionen des Straßenverkehrs in Deutschland im Zeitraum 1995 bis 2010, Umweltforschungsplan des Bundesministeriums für Umwelt, Naturschutz und Reaktorsicherheit—Luftreinhaltung-. Forschungsbericht 2008, 05 06 095 UBA-FB 96–087, Texte 73/96 (http://www.umweltbundesamt.de)
2. Landesamt für Natur, Umwelt und Verbraucherschutz (LANUV) Nordrhein-Westfalen. Jahreskenngrößen 2007. (http://www.lanuv.nrw.de/luft/immissionen/ber_tred/berichte.htm) 2008

3. Umweltbundesamt: Luft und Luftreinhaltung. Luftbelastung Deutschland: Luftbelasungssituation; 2010. vorläufige Auswertung, (http://www.umweltbundesamt.de/luft/schadstoffe/luftbelastung.htm) 2010
4. Chaloulakou, A., Mavroidus, I., Gavriil, I., Compliance with the annual NO_2 air quality standard in Athens. Required NO_X levels and expected health implications. Atmos. Environ. **42**, 454–465 (2008)
5. Anttila, P., Tuovinen, J.-P., Niemi, J.V.: Primary NO_2 emissions and their role in the development of NO_2 concentrations in a traffic environment. Atmos. Environ. **45**, 986–992 (2011)
6. Schmitzhofer, R., Beauchamp, J., Dunkl, J., Wisthaler, A., Weber, A., Hansel, A.: Long-term measurements of CO, NO, NO_2, benzene, toluene and PM10 at a motorway location in an Austrian valley. Atmos. Environ. **42**, 1012–1024 (2008)
7. Williams, M.L., Carslaw, D.C.: New directions: science and policy—out of step on NO_X and NO_2? Atmos. Environ. **45**, 3911–3912 (2011)
8. Seinfeld, J.H., Pandis, S.N.: Atmospheric chemistry and physics: from air pollution to climate change. Wiley, New York (1998)
9. Kurtenbach, R., Kleffmann, J., Niedojadlo, A., Wiesen, P.: Primary NO_2 emissions and their impact on air quality in traffic environments in Germany. Environ. Sci. Eur. **24**, 21 (2012). doi:10.1186/2190-4715-24-21
10. Clapp, L.J., Jenkin, M.E.: Analysis of the relationship between ambient levels of O_3, NO_2 and NO as a function of NO_X in UK. Atmos. Environ. **35**, 6391–6405 (2001)
11. Atkinson, R., Baulch D.L., Cox, R.A., Crowley, J.N., Hampson, Jr R.F., Kerr, J.A., Hynes, R.G., Jenkin, M.E., Kerr, J.A., Rossi, M.J., Troe, J.: In: Summary of Evaluated Kinetic and Photochemical Data for Atmospheric Chemistry, IUPAC Subcommittee on Gas Kinetics Data Evaluation for Atmospheric Chemistry, University of Cambridge, UK, www.iupac-kinetic.ch.cam.ac.uk (2005)
12. Rabl, P., Scholz, W.: Wechselbeziehung zwischen Stickstoffoxid- und Ozon-Immissionen, Datenanalysen aus Baden-Württemberg und Bayern 1990–2003. Immissionsschutz **1**, 21–25 (2005)
13. Palmgren, F., Berkowicz, R., Ketzel, M., Winther, M.: Elevated NO_2 pollution in copenhagen due to direct emission of NO_2 from road traffic. Poster presented at 2nd ACCENT symposium. www.air.dmu.dk (2007)
14. Grice, S., Stedman, J., Kent, A., Hobson, M., Norris, J., Abbott, J., Cooke, S.: Recent trends and projections of primary NO_2 emissions in Europe. Atmos. Environ. **43**, 2154–2167 (2009)
15. Fujishima, A., Honda, K.: Electrochemical photolysis of water at a semiconductor electrode. Nature **238**, 37–38 (1972)
16. Fujishima, A.: Discovery and applications of photocatalysis—creating a comfortable future by making use of light energy. Jpn. Nanonet Bull. **44**, 1–3 (2005)
17. Fracer, L.: Titanium dioxide: environmental white knight? Environ. Health Perspect. **109**(4), A174–A177 (2001)
18. Linsebigler, A.L., Lu, G., Yates Jr, J.T.: Photocatalysis on TiO_2 surfaces: principles mechanisms, and selected results. Chem. Rev. **95**, 735–758 (1995)
19. Laufs, S., Burgeth, G., Duttlinger, W., Kurtenbach, R., Maban, M., Thomas, C., Wiesen, P., Kleffmann, J.: Conversion of nitrogen oxides on commercial photocatalytic dispersion paints. Atmos. Environ. **44**, 2341–2349 (2010)
20. European PICADA Project, GROWTH Project GRD1-2001-40449, February 2006, http://www.picada-project.com/domino/SitePicada/Picada.nsf?OpenDataBase
21. Maggos, Th, Plassais, A., Bartzis, J.G., Vasilakos, Ch., Moussiopoulos, N., Bonafous, L.: Photocatalytic degradation of NO_X in a pilot street canyon configuration using TiO_2-mortar panels. Environ. Monit. Assess. **136**, 35–44 (2008)
22. Langridge, J.M., Gustafsson, R.J., Griffiths, P.T., Cox, R.A., Lambert, R.M., Jones, R.L.: Solar driven nitrous acid formation on building material surfaces containing titanium dioxide: a concern for air quality in urban areas? Atmos. Environ. **43**, 5128–5131 (2009)
23. Landesamt für Natur, Umwelt und Naturschutz Nordrhein-Westfalen, Auswirkung der Umweltzone Köln auf die Luftqualität - Auswertung der Messdaten, 2009. http://www.lanuv.nrw.de/luft/pdf/Umweltzone_Koeln_20090625.pdf

Chapter 3
Physical and Chemical Processes in the Upper Troposphere and Lower Stratosphere

Martin Riese

The upper troposphere and lower stratosphere (UTLS) is an important factor in the climate system. Particularly at the tropopause, even minor changes in components relevant for radiation, such as water, ozone, and cirrus clouds, cause significant changes in radiative forcing. The Institute for Energy and Climate Research (IEK-7) at Forschungszentrum Jülich studies the relevant processes such as the microphysics of clouds in the UTLS on local, regional and finally global scales, and also the distribution of climate-relevant trace substances in this altitude region. This involves a combination of aircraft measurements with high-resolution in situ instruments, remote sensing techniques on the ground, on aircraft and on satellites, laboratory experiments as well as simulations on several scales. The analyses stretch from the tropics and the mid-latitudes to the polar region.

In spite of its immense importance, the UTLS region is one of the least understood regions of the atmosphere. Structure and composition of the UTLS are the result of a complex interplay of various atmospheric processes that operate at different temporal and spatial scales (see Fig. 3.1).

Quantifying the processes that control UTLS composition (e.g. stratosphere-troposphere exchange) is essential for improved climate projections. Riese et al. [2] assessed the influence of uncertainties in the atmospheric mixing strength on global UTLS distributions of greenhouse gases (water vapour, ozone, methane, and nitrous oxide) and associated radiative effects. The assessment is based on multi-annual simulations with the Chemical Lagrangian Model of the Stratosphere (ClaMS) driven by ERA-Interim meteorological data and on a state of the art radiance code. The atmospheric mixing strength was varied within current uncertainty limits. Corresponding radiative effects are found to be rather large for ozone and water vapour, both characterized by steep mixing ratio gradients in the UTLS (Figs. 3.2 and 3.3).

M. Riese (✉)
Institute for Energy and Climate Research, IEK-7, Research Centre Jülich,
Jülich, Germany
e-mail: m.riese@fz-juelich.de

R. Koppmann (ed.), *Atmospheric Research From Different Perspectives*,
The Reacting Atmosphere 1, DOI: 10.1007/978-3-319-06495-6_3,
© Springer International Publishing Switzerland 2014

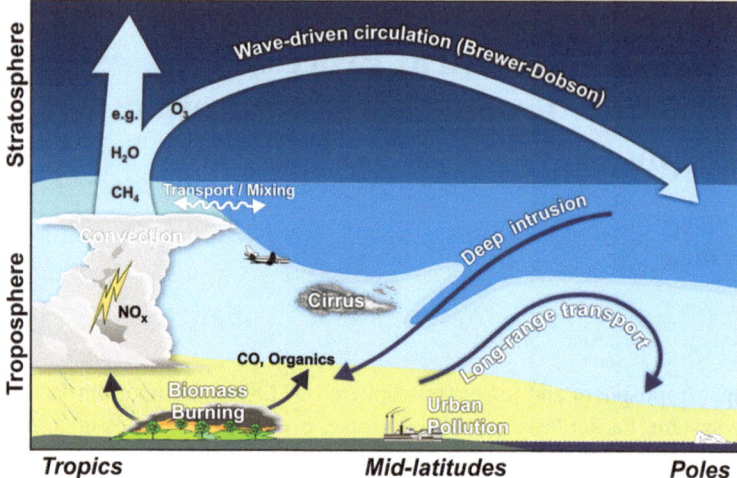

Fig. 3.1 The global structure of the troposphere and lower stratosphere. Shown are processes that determine UTLS composition. The middle shaded *blue* indicates the lowermost stratosphere, where isentropic surfaces cross the tropopause and facilitate quasi-horizontal transport, illustrated by white wave-like arrows (Figure adapted from [1])

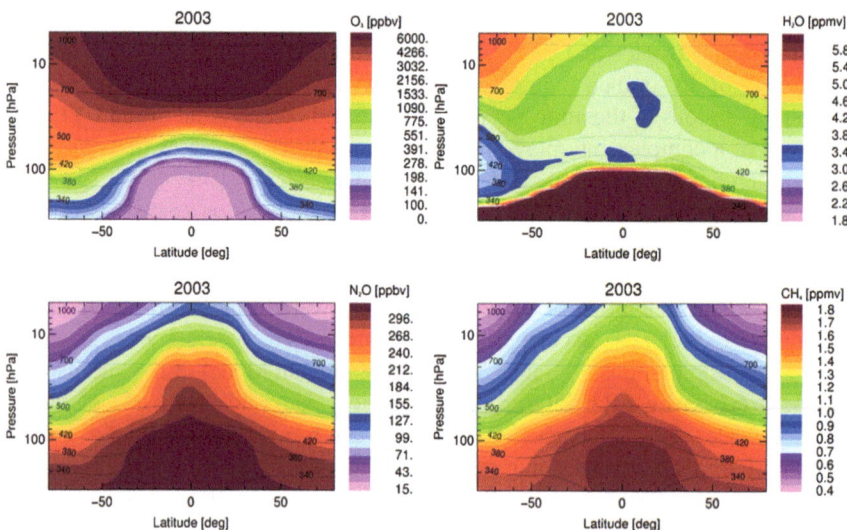

Fig. 3.2 Zonally averaged values of ozone, water vapour, nitrous oxide, and methane for 2003 taken from a multi-annual reference simulation with the CLaMS model. Pressure is used as the vertical co-ordinate (Figure adapted from [2])

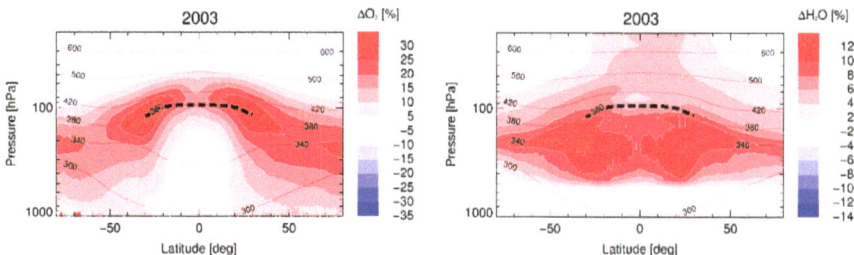

Fig. 3.3 Influence of uncertainties in the atmospheric mixing strength on simulated UTLS ozone (*left*) and water vapor (*right*). Shown are percentage differences for zonally averaged values (*2003*) obtained for two simulations with the Chemical Lagrangian Model of the Stratosphere (CLaMS), spanning the current uncertainty range of atmospheric mixing strength. For details see [2]. Corresponding globally averaged radiative effects are about 0.72 and 0.17 W/m^2 for water vapour and ozone, respectively (Figure adapted from [1])

Water Vapour Budget in the UTLS

This section summarizes a few recent observational findings related to the water budget in the UTLS.

While climate models successfully simulate the effects of water vapour on the climate in the troposphere, the simulation of key processes that determine the water vapour distribution in the stratosphere are still inadequate. This applies, in particular, to the transport of water vapour from the tropical troposphere to the stratosphere and the accompanying drying processes. Even state-of-the-art models are thus incapable of describing the variations of stratospheric water vapour observed during the last few decades. This is a critical issue since recent analyses (e.g. [3]) emphasize that changes in atmospheric water vapour are an important trigger of decadal changes in surface climate.

Atmospheric processes controlling the water vapor budget in the tropical and extra-tropical UTLS are discussed in detail in [4, 5]. Measurements carried out as part of campaigns in tropical zones of Australia (SCOUT-O3), Brazil (TROCCI-NOX) and Africa (AMMA) have shown that changes in the water vapour content in the stratosphere are closely related to changes of water vapour transport through the tropical tropopause layer [6]. This raised the question of the paths along which air masses move from the troposphere into the stratosphere and of the processes governing the drying of humid tropospheric air before it enters the stratosphere. Water input into the stratosphere is the overall result of very complex interactions between deep convection (up to an altitude of approx. 15 km), a slow diabatic ascent through the tropical tropopause layer (TTL) and the movement of these air masses towards the ascending branch of the wave-driven Brewer-Dobson circulation. According to one widely accepted theory, the air masses travel very large horizontal distances (circling the earth several times) while slowly traversing the TTL (time scale of weeks) and are "freeze-dried" in the coldest regions they pass through before entering the stratosphere. Trajectory calculations agree well with this theory and show that the temperature in the coldest region an air parcel passes through is the most important factor determining the amount of water vapour it contains before it enters the stratosphere. The drying process itself is closely related to the formation of cold cirrus clouds.

While results of recently published studies suggest that these clouds form close to the TTL only if the water vapour saturation is very high ("supersaturation puzzle"), our measurements are in line with current theories. The results of these recent studies are therefore evidently due to the fact that the measuring instruments used were not accurate enough. The results also show that, in contrast to the conclusions of several other studies, the role of overshooting convection is of secondary importance for water vapour transport into the stratosphere.

Apart from the processes described here, the oxidation of methane and molecular hydrogen in the stratosphere also contributes to the water vapour budget. IEK-7 has made a decisive contribution to quantifying this effect, primarily by analysing a long time series of balloon measurements that is unique in the world. These measurements started in Jülich in 1978 and are being continued today in cooperation with Frankfurt University [7]. The observations indicate, for example, that methane oxidation contributes less than 30 % to a 1 % per year increase of stratospheric water vapour observed between 1978 and 1998 over Boulder, Colorado [8]. Predictions for 2050 indicate that "likely increases" of tropospheric methane levels will lead to a systematic increase of upper stratospheric water vapour of 0.4 ppmv [9]. A similar value is predicted as an upper limit of effects of a future hydrogen economy (see below).

The potential impact of a future hydrogen economy on the development of stratospheric water vapour as well as possible effects on the ozone layer and climate is analysed in the review paper by [10]. Several studies highlighted the positive impacts of a hydrogen economy on air quality, climate and health caused by the reduction of emissions from fossil fuels such as greenhouse gases and many pollutants such as nitrogen oxides. The expected benefit depends critically on the method used for production and storage of hydrogen, in particular, on renewable energy sources such as wind-powered electrolysis. Some studies discussed a negative impact on our environment caused by a systematic increase of the water vapour concentrations in the stratosphere, in particular at high latitudes. Vogel et al. show, however, that this increase and its impact on both stratospheric cooling and stratospheric polar ozone loss is comparably low compared to the variability of stratospheric water vapour values [10]. Provided that hydrogen is produced from renewable energy sources, the environmental benefits and the minor risks for the stratosphere therefore reinforce the conclusion that hydrogen as an energy carrier is a reasonable alternative to fossil fuels.

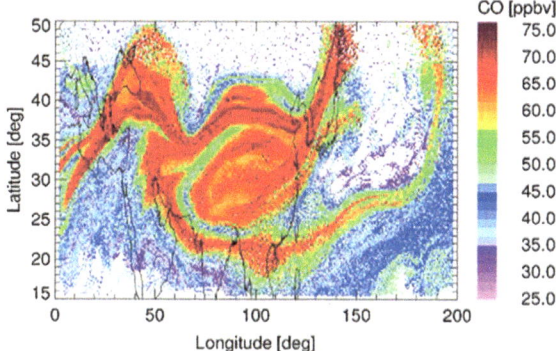

Fig. 3.4 CO distribution (right panel) in the upper part of the Asian monsoon anti-cyclone ($\sim$18 km) for August 9, 2003. The Asian monsoon has a strong influence on the composition of the tropical tropopause region (TTL). "Young" tropospheric air (high CO values) result from fast convective upward-transport in the centre of the Asian monsoon. Quasi-horizontal mixing of "older" extra-tropical stratospheric air (low CO values) into the TTL occurs at the edge of the highly variable anticyclone

Exchange Between Troposphere and Stratosphere in the Asian Monsoon Region

Studies on the impact of the energy industry (e.g. emission of greenhouse gases) on the upper troposphere and lower stratosphere (UTLS) are particularly important in the context of the research network "reacting atmosphere". The great population growth in Southeast Asia and the accompanying rapid economic growth cause increasing regional and global pollution of the atmosphere. The Asian summer monsoon in particular constitutes a persistent circulation pattern transporting climate-relevant emissions from the surface boundary layer to the lower stratosphere. In order to sustainably reduce the related global effects on the climate, it is very important to understand such transport paths in detail [11]. In addition, the impact of climate change on such processes, in particular on the regional scale, is only partially understood. The question of exactly how the emissions from Asia and their secondary products enter the stratosphere also remains unanswered.

In the upper troposphere, the Asian monsoon forms a high-pressure area at an altitude of approx. 10–18 km that is almost stationary for 3 months and covers almost all of Asia in the form of an anticyclone. This anticyclone, with its core located at approx. 30° N, very effectively traps emissions from Asia transported to this altitude by a strong convection current, particularly above India, southern China and Indonesia. As a direct consequence, elevated levels of tropospheric trace gases such as water vapour, CO (Fig. 3.4) or HCN can be found in the anticyclone, as well as reduced amounts of stratospheric trace gases such as ozone or HCl. This finding, which has been the subject of much discussion in the last few years, was derived mainly from satellite observation data [12]. However, the extent to which emissions from Asia influence the composition of the tropical tropopause layer (TTL) is still uncertain. The TTL is a region of the atmosphere stretching across the equator which has a decisive impact on mass transport into the stratosphere and is therefore referred to as the "gateway to the stratosphere". Meridional transport from the anticyclone into the TTL in the direction of the equator is intermittent and strongly influenced by planetary waves and mixing processes. Current research aims to make qualitative statements regarding the amounts of emissions transported.

We studied the impact of this circulation pattern on the annual cycle of ozone in depth as part of model simulations [13, 14]. It was also discussed how the representation of vertical transport in models influences the composition of the TTL [15]. The contribution of peat fires in Indonesia to the elevated HCN levels in the TTL was verified by means of the CLaMS model [16]. All these studies do not only show that the Asian monsoon is an important additional transport path connecting the emission sources in Southeast Asia with the stratosphere, but also that on this path a significant part of the stratospheric air masses from the extra-tropics is mixed into the TTL. These and other open questions need to be addressed in the near future on the basis of high-flying research aircrafts (e.g. Geophysica, ER-2), equipped with sophisticated instrumentation.

References

1. Riese M., Oelhaf, H., Preusse, P., Blank, J., Ern, M., Friedl-Vallon, F., Fischer, H., Guggenmoser, T., Höpfner, M., Hoor, P., Kaufmann, M., Orphal, J., Plöger, F., Spang, R., Suminska-Ebersoldt, O., Ungermann, J., Vogel, B., Woiwode W.: Gimballed Limb Observer for Radiance Imaging of the Atmosphere (GLORIA) scientific objectives. Submitted to Atmospheric Measurement Techniques (2014)
2. Riese, M., Ploeger, F., Rap, A., Vogel, B., Konopka, P., Dameris, M., Forster, P.M.: Impact of uncertainties in atmospheric mixing on simulated UTLS composition and related radiative effects. J. Geophys. Res. (2012). doi:10.1029/2012JD017751
3. Solomon, S., Rosenlof, K.H., Portman, R.W., Daniel, J.S., Davis, S.M., Sanford, T.J., Plattner, G.-K.: Contributions of stratospheric water vapor to decadal changes in the rate of global warming. Science **327**, 1219–1223 (2010)
4. Fueglistaler, S., Dessler, A.E., Dunkerton, T.J., Folkins, I., Fu, Q., Mote, P.W.: Tropical tropopause layer. Rev. Geophys. 47 (2009). doi: 10.1029/2008RG0000267
5. Gettelman, A., Hoor, P., Pan, L.L., Randel, W.J., Hegglin, M.I., Birner, T.: The extratropical upper troposphere and lower stratosphere. Rev. Geophys. 49 (2011). doi: 10.1029/2011RG00035
6. Schiller, C., Grooß, J.-U., Konopka, P., Plöger, F., Silva dos Santos, F.H., Spelten, N.: Hydration and dehydration at the tropical tropopause. Atmos. Chem. Phys. **9**, 9647–9660 (2009)
7. Rohs, S., Schiller, C., Riese, M., Engel, A., Schmidt, U., Wetter, T., Levin, I., Nakazawa, T., Aoki, S.: Long-term changes of methane and hydrogen in the stratosphere in the period 1978-2003 and their impact on the abundance of stratospheric water vapor. J. Geophys. Res. **111**, D14315 (2006). doi:10.1029/2005JD006877
8. Oltmans, S.J., Vömel, H., Hofmann, D.J., Rosenlof, K.H., Kley, D.: The increase in stratospheric water vapor from balloonborne, frostpoint hygrometer measurements. Geophys. Res. Lett. **27**(21), 3453–3456 (2000). doi:10.1029/2000GL012133

9. Riese, M., Grooß, J.-U., Feck, T., Rohs, S.: Long-term changes of hydrogen-containing species in the stratosphere. J. Atmos. Solar Terr. Phys. **68**(17), 1973–1979 (2006)
10. Vogel, B., Feck, T., Grooß, J.-U., Riese, M.: Impact of a possible future global hydrogen economy on Artic stratospheric ozone loss (mini review). Energy Environ. Sci. **5**(4), 6445–6452 (2012). doi:10.1039/c2ee03181g
11. Randel, W.J., Park, M., Emmons, L., Kinnison, D., Bernath, P., Walker, K. A., Boone, C., Pumphrey, H.: Asian monsoon transport of pollution to the stratosphere. Science 328, 611–613 (2010). doi: 10.1126/science.1182274
12. Park, M., Randel,W.J., Gettelman, A., Massie, S.T., Jiang, J.H.: Transport above the Asian summer monsoon anticyclone inferred from Aura Microwave Limb Sounder tracers. J. Geophys. Res.-Atmos. 112, D16309 (2007). doi: 10.1029/2006JD008294
13. Konopka, P., Grooß, J.-U., Günther, G., Ploeger, F., Pommrich, R., Müller, R., Livesey, N.: Annual cycle of ozone at and above the tropical tropopause: observations versus simulations with the Chemical Lagrangian Model of the Stratosphere (CLaMS). Atmos. Chem. Phys. **10**, 121–132 (2010). doi:10.5194/acp-10-121-2010
14. Ploeger, F., Konopka, P., Günther, G., Grooß, J.-U., Müller, R.: Impact of the vertical velocity scheme on modeling transport in the tropical tropopause layer. J. Geophys. Res. **115**, D03301 (2010). doi:10.1029/2009JD012023
15. Ploeger, F., Fueglistaler, S., Grooß, J.-U., Günther, G., Konopka, P., Liu, Y.S., Müller, R., Ravegnani, F., Schiller, C., Ulanovski, A., Riese, M.: Insight from ozone and water vapour on transport in the tropical tropopause layer (TTL). Atmos. Chem. Phys. **11**, 407–419 (2011). doi:10.5194/acp-11-407-2011
16. Pommrich, R., Müller, R., Grooß, J.-U., Günther, G., Konopka, P., Riese, M., Heil, A., Schultz, M., Pumphrey, H.-C., Walker, K. A.: What causes the irregular cycle of the atmospheric tape recorder signal in HCN? Geophys. Res. Lett. **37**, L16805 (2010). doi:10. 1029/2010GL044056

Chapter 4
Modelling and Numerical Simulation

Matthias Ehrhardt, Michael Günther and Birgit Jacob

In this field of the research network we are concerned with the development of methods based on mathematical modelling, system theory and numerical analysis, in order to analyse, improve and simulate numerically in an efficient and robust way the system reacting atmosphere on sub and overall model level. A transdisciplinary research approach is inevitable, combining modelling and simulation expertise of mathematics, scientific computing, atmospheric physics and chemistry, economics, and social sciences available in this research network.

Our aim is to construct a tool for illustrating the broad range of interdependencies in the overall system (humans-atmosphere-air quality-climate) and their impacts on air quality and climate. This will be used in schools or in policy consulting and shall yield quantitative guidance for future air quality policy.

The Research Questions

For many decades, the atmosphere has been changing at an increasing rate owing to changes in societal and economic developments. In turn, physical and chemical changes in the atmosphere feed back on society and economy via changes of air quality and climate. For an integrative understanding of this Overall System, one has to understand the role of the atmosphere for air quality and climate change and interactions of atmospheric change with societal and economic developments including feedbacks. This defines the key research questions of our consortium:

- What is the influence of the reacting atmosphere on society and economy via changes in air quality and climate change?

M. Ehrhardt (✉) · M. Günther · B. Jacob
Institute of Mathematical Modelling, Analysis and Computational Mathematics,
University of Wuppertal, Wuppertal, Germany
e-mail: ehrhardt@math.uni-wuppertal.de

R. Koppmann (ed.), *Atmospheric Research From Different Perspectives*,
The Reacting Atmosphere 1, DOI: 10.1007/978-3-319-06495-6_4,
© Springer International Publishing Switzerland 2014

- What is the influence of societal and economic developments on air quality and climate change via the reacting atmosphere?
- What knowledge is needed to develop sustainable strategies for improvements in air quality and climate protection?

The main task is to provide modelling and simulation methodologies needed for answering these questions. The research needed to fulfil this task will be guided by the following three key research questions:

- How can (physico-chemical) atmospheric and socio-economic models be coupled and validated to describe the Overall System?
- Which mathematical innovations in modelling, analysis and numerical simulation are needed for analysing, parameterising, simulating, coupling and applying these models to both model regions?
- How can these models be used to develop optimal measurement strategies and to measure future invention scenarios?

State-of-the-Art

In the last years, several internationally well-established models have been derived that describe the dynamics of either the physico-chemical or socio-economic components, based on two different modelling paradigms: differential equations on one side, and agent and emission flow based models on the other one. Modelling, analysis and efficient numerical simulation of coupled differential equations have been an important research topic throughout the last two decades. For coupled systems based on differential equation models (Ordinary Differential Equations, Differential-Algebraic Equations and Partial Differential Equations), monolithic and co-simulation strategies have been developed and applied to various problems, from engineering (nanoelectronic circuits) to environmental science (flood forecasting in rivers) and medicine (blood circulation in the human body). For co-simulation it turned out that the coupling structure of the joint model to be solved strongly defines the stability properties of the iterative procedure involved, which makes it possible to tune the convergence of co-simulation by appropriate modelling of the coupling structure [1].

Up to now, corresponding results are missing for heterogeneous systems linking differential equation based models to emission flow and agent based ones. However, for a holistic and thus realistic description of the overall system, coupling of the existing models and the integration of specifically developed models is mandatory to successfully analyse the coupled socio-economic and physico-chemical systems describing the overall system of the reacting atmosphere. A unique challenge is to link both model paradigms, the agent and emission flow based socio-economic and the differential equation based physico-chemical models. Only very few studies are available on proper validation strategies of such complex models.

To fill this gap, we focus on the development and application of robust and efficient models as well as algorithmic methods for simulating the Overall System in a real-time manner.

Modelling: A set of appropriate models is required to describe the key aspects of the reacting atmosphere: chemistry, transport, and radiative transfer from the regional up to the global scale for both the troposphere and the lower stratosphere considering different time scales.

The different spatial parts of the reacting atmosphere, in particular in the vertical direction, shall be modelled by different differential equation models. Consequently, different step sizes can be applied. These models have to be coupled appropriately in a numerical simulation and a proper treatment of uncertainties in physical parameters of the model by random variables has not been considered yet. Right in this context a mathematical challenge will be to incorporate properly the coupling of the physico-chemical PDE models of the atmosphere with the socio-economic models of the human input factors. Closely related to this question is the study of the sensitivities of existing sub models or the resulting overall system with respect to key parameters that may not be known or at least uncertain. From those considerations a natural task would be the development of gradings for comparison studies of atmospheric models. Furthermore, it will be of great interest to analyse ill-posed systems and their break-over points.

Numerical methods: Secondly, it is a current research issue to develop robust and efficient numerical methods to solve efficiently the large PDE system arising in atmospheric modelling. Here state-of-the-art model order reduction (MOR) methods, operator splitting and stiff multirate approaches are needed and have to be improved substantially to tackle this complex problems. Up to now, mathematical tools are not used to optimize measure points and measure campaigns in order to minimize the effort for the needed measurements. In addition, there are no established methods to assess the model sensitivity with respect to disturbances triggered by natural events or human activities.

Virtualization: The increasing complexity of such models requires a substantial increase in computing performance, which will only be possible by adapting the codes to the architecture of future generation supercomputers. Visualisation and virtualisation are important emerging tools that

- enhance scientific understanding by visualizing simulation results,
- inform decision makers about potential consequences of altered Earth System state variables, and
- attract and educate students, teachers and the general public on topics concerning the reacting atmosphere.

Presently, it is not clear how to simulate these large, complex, transdisciplinary models in a real-time manner. Besides sophisticated MOR techniques there is the need for codes which can efficiently make use of the special hardware of current high-performance computers as well as to exploit the opportunities arising from the use of a grid infrastructure.

Vision

Our vision is to obtain a whole hierarchy of mathematically founded models of growing complexity to describe the overall system, by coupling chemical, physical, economical, societal and technological sub models. These models will be critically examined using mathematical system theory and numerical analysis techniques in order to analyse, validate and extend their predictive capability on a practical and theoretical level. Furthermore, the sensitivity of the overall system with respect to a wide range of (possibly uncertain) parameters is studied in detail.

This model hierarchy extends from very detailed sub-systems up to qualitative, macroscopic models describing the overall system. By virtualization of the results, the latter models facilitate a deep qualitative and to some extent a quantitative understanding of the interdependencies of the different components. This is necessary for assessing the impact of human activity on air quality and climate on regional and global scales, leading to quantitative guidance for policy makers.

To reach our vision, we define mid-term and long-term goals and plans. On this way, we have already achieved first results.

Mid-Term Goals/Plans

The mid-term objective is to create a toolbox of analytical and numerical methods for the analysis, robust and efficient numerical simulation and real-time virtualization of both model regions. This toolbox is tailored to future exascale supercomputers and the resulting fancy, easy accessible simulator for qualitative simulation of the reacting atmosphere is used for showing possible impacts on the air quality.

To cope with the complexity, heterogeneous temporal and spatial scales and inter-woven nature of these model regions, key atmospheric models will be re-engineered to exploit new mathematical methods. For that purpose we will classify the different PDE models, specify appropriate boundary conditions that lead to well-posed sub-models and present construction of coupling approaches of the different sub models of existing simulators. In order to properly obtain submodels we investigate in detail operator splitting methods that separate the different physical, chemical and socio-economic submodels in each time step. Here we develop operator splitting methods for non-autonomous stiff differential equations and analyse concisely the induced splitting error. To cope with uncertainties we present a strategy to select random parameters in state-of-the-art models as well as possibly novel models introduced in the project. This approach leads to the application of the polynomial chaos approach for solving the stochastic PDE models.

Moreover, we apply successfully (linear) MOR techniques to existing models and perform first steps to use different nonlinear MOR approaches to break down the order of the complex PDE models. We design special multiscale and multirate for simulation of the large stiff systems modelling the chemical reactions hereby improving significantly the existing IMPACT (family) approach that suffers from convergence problems. Here we pay attention to construct solvers that preserves important properties like the mass and the positivity of the concentrations exactly on the discrete level. We analyse of different continuous and discrete models with respect to stability, sensitivity and reliability.

We will develop and analyse agent-based models describing the decision-making process of the relevant stakeholder groups, as well as dynamic models for judging possible future emission paths for intervention scenarios. These models will be applied to both model regions and linked to the physico-chemical models by defining appropriate interfaces. Based on that, we will develop optimal observation strategies, i.e. the optimal choice of measurements, of the emissions of interest in the model systems, taking into account possibly diverging objectives of different stakeholder groups.

Long-Term Goals/Plans

The long-term objective is the mathematical description of the overall system and the salient interactions between the system components from chemistry, physics, economy and social science. This includes the development of a comprehensive toolbox of analytical and numerical methods for the analysis, robust and efficient numerical simulation and validation of the overall system. In case of large numbers of random parameters, application of the analysis of variance (ANOVA) approach to identify a reduced set of relevant parameters in the polynomial chaos technique.

The coupling of several subsystems of different type and complexity from chemistry, physics, economy, social science leads to a tailor made multi-math coupling theory. Doing so, we obtain as a side product a new theory of model grading with respect to reliability, stability, numerical effort, detailedness, quantification of model errors, etc. There will be different model configurations for fast-response virtualisation runs, detailed process studies and long-term impact assessment studies.

This research domain will develop new approaches to map theories and models onto Exascale systems and develop highly scalable numerical algorithms suitable for Exascale computing. With the constructed toolbox of numerical methods, a virtualization of the results of the overall system will be achieved with real-time feed- back resulting from user participation/input. The observables required for the various models, which have to be provided by measurements, will be optimised with respect to accuracy for given resources. Sensitivity analysis and uncertainty quantification will be resolved by combining the existing approaches with these specific methods. Labile systems will be considered and investigated with respect to their controllability. This knowledge will be adapted to various use cases depending on the question at hand.

First Steps

Already in the first year, some first steps have been made to start the research program, focusing first on modelling and numerical simulating the physical and chemical processes in the atmosphere.

Describing the impact of uncertainties in input parameters and processes by stochastic modelling: Currently, deterministic models are used to describe physical and chemical processes of the atmosphere. However, input parameter and processes are afflicted with small or large uncertainties, e.g., in the transport part of wind vectors, in the diffusion part of boundary layers and in the temperature of the reaction part. In addition, emission rates and initial values are uncertain meteorological data. In a joint research project, Hendrik Elbern (Research Centre Jülich) and Roland Pulch (University of Greifswald, formerly University of Wuppertal) are developing stochastic models to quantify such uncertainties by describing input data via random variables and processes: the numerical results are thus equipped with confidence intervals as error bounds. To minimize the number of random variables to be modelled, the crucial parameters have to be identified a priori, taking the level of uncertainty and impact on the solution into account. Based on a mathematically stringent development of the stochastic models, the project aims at a rigorous analysis of the impact of uncertainties, allowing for a reliable error control.

Novel mathematical approaches for estimating air quality and climate forcing emissions of technical energy processes: Climate and air quality variability are ultimately driven by emissions from technical energy conversion processes, engendering fluxes of greenhouse gases, reactive gas emissions, and aerosols into the atmosphere. However, fluxes can hardly be observed directly and reliably quantified. This fact severely hampers progress in predictive skills of climate and atmospheric chemistry models. Today's flux inversion techniques suffer from reliability. In case of greenhouse gases, even with cutting edge methods, errors in initial values directly impact the result of emission flux estimates.

By introducing novel mathematical approaches to inversion techniques [2], combining advanced models with remote sensing data, Hendrik Elbern (Research Centre Jülich) and Birgit Jacob (University of Wuppertal) dwarf this problem in a joint research project by breaking new grounds on emission flux estimation, addressing algorithmic designs of the emission inversion problem to estimate fluxes of reactive gases into the atmosphere [3]. An "Extended Ensemble Kalman Smoother" with full adjoint modelling will be developed, relying in part on the existing adjoint chemistry transport model EURAD-IM.

Modelling and numerically simulating the pollutant transport and diffusion in the atmosphere: The process of pollutant transport and diffusion in the atmosphere (and in water) is described by an advection-diffusion equations with given parameters, like the components of the wind velocity, the falling velocity of the pollutants by gravity, the power of the source, the transformation coefficient of pollutants and the horizontal and vertical diffusion coefficients.

Numerical methods to solve this type of equations must conserve the positivity of the computed concentration. In the work [4] we are involved from the mathematical point of view with qualitative characteristics of difference schemes and investigation of positivity of some difference schemes for advection-diffusion equations. Also, these specially designed difference schemes must be accompanied with artificial boundary conditions to limit the computational domain appropriately, since the original geographical domain is often too large. If the solution on the bounded computational domain coincides with the solution on the unbounded half-space (restricted to the bounded domain) one refers to these conditions as transparent boundary conditions (TBC). Finally, these boundary conditions are designed on a discrete level directly for the chosen numerical schemes in order to conserve the monotonicity property and to prevent any unphysical reflections at these boundaries.

In a second paper [5] we considered the problems of optimal location of industrial enterprises and optimization of emissions from enterprises for ensuring sanitary environment criteria. Moreover, we studied the problem of determination of the diffusion coefficients and the coefficient of transformation of aerosols.

For the accurate numerical approximation of sources one needs appropriate approximations of the delta function. Especially, for the method of lines one needs unsymmetric approximations that, up to the author's knowledge, do not exist in the literature. Recently, in [6] we proposed new delta-convergent sequences that vanish at the support of the limit Dirac delta function. These are sequences of even functions that don't have a compact support. We developed some results concerning delta sequences and stated examples of delta sequences of the type with or without compact support being even or not even.

Investigating the chemical reaction kinetics implemented in IMPACT: In a master thesis, jointly supervised by Jens-Uwe Grooß (Research Centre Jülich) and Michael Günther (University of Wuppertal), David Scheinmann has carried out a rigorous analysis of the steady-state and family approach used within IMPACT to model the reaction kinetics of chemical species in the atmosphere [7]. Both approaches have turned out to be equivalent to semi-explicit DAE models. Based on this interpretation and a series of numerical experiments, the divergence of IMPACT in some settings turned out to be caused by an inappropriate modelling of the NO_x family.

This interpretation has opened the door for using DAE methods and incorporating multirate and co-simulation techniques in the numerical simulation of the chemical reaction kinetics [8].

References

1. Bartel, A., Brunk, M., Günther, M., Schöps, S.: Dynamic iteration for coupled problems of electric circuits and distributed devices. SIAM J. Sci. Comput. **35**, B315–B335 (2013)
2. Elbern, H., Strunk, A., Schmidt, H., Talagrand, O.: Emission rate and chemical state estimation by 4-dimensional variational inversion. Atmos. Chem. Phys. **7**, 3749–3769 (2007)

3. Wu, X., Jacob, B., Elbern, H.: Location analysis of observations for atmospheric chemical transport model with emissions. Submitted for Publication
4. Dang, Q.A., Ehrhardt, M.: Adequate numerical solution of air pollution problems by positive difference schemes on unbounded domains. Math. Comput. Model. **44**, 834–856 (2006)
5. Dang, Q. A., Ehrhardt, M., Tran, G. L., Le, D.: On the numerical solution of some problems of environmental pollution, Chapter 6. In: Robinson, P., Gallo, R. (eds.) Environmental Chemistry Research Progress, vol. 11788, pp. 199–228. Nova Science Publishers, Inc., Hauppauge (2009)
6. Dang, Q.A., Ehrhardt, M.: On Dirac delta sequences and their generating functions. Appl. Math. Lett. **25**(12), 2385–2390 (2012)
7. Scheinmann, D.: ASAD: Eine kritische Analyse der mathematischen Modellierung und numerischen Simulation der Reaktionskinetik in der Atmosphärenchemie. Master Thesis, Bergische Universität Wuppertal (2012)
8. Striebel, M., Bartel, A., Günther, M.: A multirate ROW-scheme for index-1 network equations. Appl. Numer. Math. **59**, 800–814 (2009)

Chapter 5
Sustainable Strategies

Manfred Fischedick

Role in the Research Network and State of the Art

Our research focuses on socio-economic process understanding and impact analysis. It closes the circle between the changing intrinsic conditions in the reacting atmosphere (mainly described in the other research topics), the resulting needs for action and if specific policy targets regarding air quality and/or climate protection are pursued, and corresponding divergent emissions paths, which again feed back to the reacting atmosphere. Particularly, we aim at the development of a holistic and integrated policy approach, which is able to trigger the resulting needs for action in regions of relevance.

Based on detailed knowledge of the complex decision making processes in policy, industry, economy, society, suitable technological and non-technological measures and instruments as well as policies will be identified and systematically assessed, considering the full set of socio-economic parameters. The formulation of an integrative air quality and climate policy concept considering the complex decision making processes and reflecting the full set of technological and non-technological options is the key goal of the proposed research. Suitable intervention scenarios will have been developed and widely discussed as well as disseminated amongst policy decision makers based on the further development of existing modelling approaches.

Emission pathways significantly determine the reactions and the behaviour of the atmosphere. So far, particularly in the context of greenhouse gas emissions (GHG emissions) several emission inventories and assessment models have been developed for different regions (cf. [1]). Furthermore, scenario tools are available at least for some world regions presenting assumptions or projections (scenarios) for the future development of emissions considering future structural changes on the basis of a set of underlying measures including policy instruments (cf. [2]).

M. Fischedick (✉)
Wuppertal Institute for Climate, Environment and Energy, Wuppertal, Germany
e-mail: manfred.fischedick@wupperinst.org

R. Koppmann (ed.), *Atmospheric Research From Different Perspectives*,
The Reacting Atmosphere 1, DOI: 10.1007/978-3-319-06495-6_5,
© Springer International Publishing Switzerland 2014

However, these models mostly focus on technological questions, exclude socio-economic aspects and do not follow a cross-sectoral and cross-problem perspective (cf. [3]). In particular the latter aspect is true for the combination of air quality and climate change as two major challenges for the future, especially in the East Asian Megacities (EAM).

Actually, with the Greenhouse gas—Air Pollution Interaction and Synergies (GAINS) model from International Institute for Applied Systems Analysis (IIASA) and the IES (Integrated Environmental Strategies) model from the U.S. Environmental Protection Agency only two models exist combining both aspects in an integrated approach. At least for some countries in different world regions (e.g. EU, USA, India, China) these models enable an analysis of strategies that maximize co-benefits between air pollution control and GHG mitigation and indicate how to avoid trade-offs. However, also these models mainly concentrate on technology and on cost aspect (reduction of abatement costs through combined strategies), but do not specifically focus on other socio-economic aspects (e.g. social implications, employment aspects, public perception) being critical for the implementation of the analysed strategies. Moreover, as these models follow a more aggregated approach, they do not specifically address the city or regional level where socio-economic effects become manifest.

In the fast-growing East Asian Megacities (EAM), which are of specific interest for the whole research network, the air pollution situation is extremely serious. The corresponding protection of human health is urgent and determines the policy action. Considering the fast economic growth leading to an increasing demand for fossil fuels, from a global perspective EAM are confronted with a strong request for implementation of climate protection measures. It is generally understood that air pollution control mechanisms can lead to positive or negative effects in terms of climate protection. While the reduction of black carbon (e.g. through a change of the energy mix) or ozone precursor emissions directly reduces anthropogenic radiative forcing, most of the end-of-the-pipe technologies aiming for air pollution control are accompanied with energy efficiency losses and therefore increasing GHG emissions. On the other hand many of active climate protection measures (e.g. energy efficiency improvements, reduction of the number of coal fired power plants) can considerably contribute to a reduction of air pollution, those kinds of benefits are often overlooked. Contrary, ill-designed climate protection options could counteract air quality efforts (e.g. combustion of biomass in small household applications could lead to a significant increase of particle emissions).

In that context, a more holistic approach, a combination and integrative judgement of sectoral findings and strategies including the whole area of socio-economic aspects seems to be necessary, an approach that considers the interdependencies between dedicated technical measures as well as between the complex decision making process in industry, policy and society related to climate protection and air quality improvements. The research topic "sustainable strategies" is going along this way and pursues to develop integrated strategies (intervention scenarios) as an appropriate answer regarding the identified system interdependencies and resulting needs for (additional or modified) actions considering specific climate protection and/or air quality goals in a selected research region.

Thus, the research in this context is constitutional for the bottom and left part of the overall system described before (cf. Chap. 1). In fact, it delivers the basis for a dynamic set of input parameters for the examination of the reacting atmosphere, which so far mostly deals with a "status quo approach" of emissions assumptions. Moreover, as the project as a whole tries to close the complex cycle (overall system approach), an iterative process will be started. The intervention scenarios will change the quality and the quantity of emissions, as a consequence atmospheric processes might to be influenced. The resulting impacts on air quality and climate impact have to be compared with the original intention of the dedicated intervention scenarios and the scenarios have to be adapted accordingly if there are any significant differences.

In terms of the specification of intervention scenarios we focus on EAM, but make use of the experience of other world regions, particularly in Ruhr Metropolitan Region (RMR). Having the learning from the past in mind and based on a deeper understanding of the decision making process, for EAM it will be assessed how a consistent set of measures (mix of technological, political and socio-economic measures) could look like, how the identified strategies (described in scenarios) could be implemented successfully and which kind of positive impacts might emerge (e.g. entrepreneurial opportunities) or trade-offs have to be solved. An in-depth understanding of the decision making processes (in policy, industry, economy, society) is absolutely crucial for this working step and constitutional for the discussion how the corresponding results of the research work can be communicated to the relevant stakeholders and can be transferred to other regions.

With respect to the aforementioned goals the key research questions are:

- How do climate protection efforts interact with air quality improvement measures and vice versa?
- How can a suitable mix of local/regional measures comprising technological, political and socio-economic options aiming for an integrated air quality and climate protection approach look like?
- How can an integrative local/regional strategy be implemented and a corresponding transformation process be shaped?

In terms of the climate policy aspect and corresponding measures it has to be considered that various policy levels are relevant (local, urban, national and international) in the decision making process, mostly different policy institutions are responsible for the implementation of specific policy measures. For instance, nowadays the European Commission is responsible for a huge number of mandatory emission standards and rules in the member states (e.g. in terms of concentration limitations for SO_2, NO_2 and PM10). The implementation of a successful strategy necessarily requires an integrated policy across different policy arenas (multi-level policy approach). An additional guiding question is therefore:

- How can a regional strategy be realized as consistent part of a multi-level policy mix?

Focussing on the specific model region (EAM), the vision is to obtain an integrative air quality and climate strategy by knowing all interdependencies within the Overall System and a deep understanding of the decision making processes. For EAM this will not be done from the scratch, but through making use of experience from successfully implemented transformation processes in the past. Particularly experience and learnings from the Ruhr Metropolitan Region (RMR) can be used.

Thus, the mid-term objective is to get a better understanding of system behaviour and preconditions for successful transformation processes based on the historical experience in different areas as well as promising future concepts. Relevant frame conditions such as already fixed policy targets regarding air quality and climate protection, already implemented policies and instruments, expected policies, mid- and long-term goals will be compiled.

Suitable technical and non-technical measures in both fields of air quality and climate protection will be collected and systematically assessed on the basis of a transparent and comprehensive set of criteria (multi-criteria assessment). Implications for economy and society will be discussed in depth and resulting constraints and opportunities for the practical implementation will be derived. With regard to economy, resulting challenges for business companies will be analysed, e.g. entrepreneurial opportunities, innovation needs, business portfolio, incentive schemes. Furthermore, the issue of overall system compatibility will be addressed as a criterion to avoid problem shifting (negative impacts on other relevant environmental areas beyond the atmosphere, e.g. loss of biodiversity, land use change).

In cooperation with the other research areas a dynamic modelling tool will be developed to identify and systematically judge possible future emission paths bundled in consistent intervention scenarios. Beside these emission models, an agent based modelling tool will be developed that helps to reflect the decision making process of the important stakeholder groups. This will be based on empirical data or scientific based hypothesis for stakeholder behaviour in relevant fields of policy, industry, economy and society.

Considering the long history of the air quality improvements in the RMR, successful air quality measures and corresponding policy tools will be analysed, particularly in terms of their transferability for relevant regions. It will be discussed how this kind of experience can be adapted to EAM and might help to accelerate technological progress and implementation. Particularly experience from the ongoing Innovation City process, which aims for a 50 % reduction of greenhouse gas emissions between 2010 and 2020 in a specific city of RMR, can serve as a major analytical basis. This is a unique opportunity as the Innovation City can be used as a "laboratory" to test and improve ideas and concepts developed in this project.

The objectives can only be reached through close interdisciplinary cooperation within the whole research network. Already existing collaborations, among others the formal cooperation of one of the network partners with the Tsinghua University of Beijing, shall be extended and intensified and additional project based cooperation, particularly with Asian partners, shall be established to cover the specific cultural and policy framework. The ongoing close cooperation with the Innovation City Ruhr will be continued.

Project Experience, Methodological Background and On-going Research

The intended research of the research network "The Reacting Atmosphere" directly ties in with the experience from past and ongoing research work of the different network partners. Particularly it very well fits with the research agenda of the Wuppertal Institute. Research reflecting the development of cities and regions towards sustainability is for different reasons an exemplary integrative research field and can perfectly be related to the transition concept. Consequently, the transition cycle serves as structuring element for the research of the institute in this area.

In order to understand in greater depth the dynamics and patterns of transition processes, the "multi-level concept" builds the central concept of the transition research approach. Transitions—defined as a "radical, structural change of a societal (sub)system that is the result of a co-evolution of economic, cultural, technological, ecological and institutional developments at different scale-levels" [4]—generally take place due to interaction between and interconnected developments on three different functional scale-levels. According to the authors of the Dutch KSI-network, structures are found on three different scale levels conceptualized as follows ([5]; cf. Fig. 5.1).

First, the so-called "socio-technical landscape"—characterized by either very slow changes that can hardly be influenced (e.g. climate change, industrialization processes, globalisation and urbanization trends) or external shocks, like wars [5]. The structure on the landscape level thereby builds the overarching basic conditions for the other two functional levels.

The second level is the so-called "socio-technical regime". These socio-technical regimes are characterized by cognitive (belief systems, guiding principles, goals, innovation agendas, problem definitions, search heuristics), regulative (standards, laws) and normative (values, roles, behavioural norms) rules shared by the relevant actors of a regime (governments, companies, civil society, scientists, etc.). The regime builds up the dominant structure, culture and practices—i.e. the centre of power—within a system.

The third level is the so-called "socio-technical niche" composed of individual actors, technologies and practices, from which radical innovations and frontrunner movements emerge. Some of these elements have the power and dynamic to trigger a transformation process and to change specific socio-technical regimes.

Fig. 5.1 Interaction between landscape, regime and niches (*Source* [5])

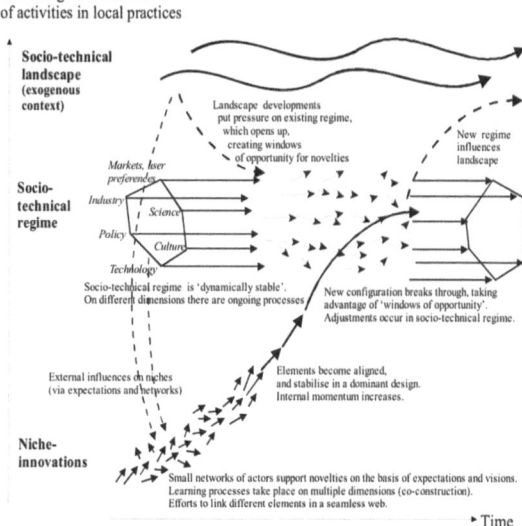

Figure 1: Multi-level perspective on transitions (adapted from Geels, 2002: 1263)

By means of the multi-level concept, the authors try to describe "the dynamics of transitions in (functional) space as the interactions between three different functional scale-levels" [5]. The necessary condition for a transition to be realized successfully is that the structural developments on each level mutually reinforce each other and point into the same direction. This applies for both evolutionary transitions (i.e. whose outcome is not planned) and goal-oriented (teleological) transitions ("in which [...] goals or visions of the state are guiding public actors and orienting the strategic decisions of private actors") [6]. Transitions may take place when there are instabilities on the regime-level, caused by tensions between the regime and its environment (either niche- or landscape-level) or learning and adaptation processes at the regime-level itself. These tensions create windows of opportunity for a niche to become more powerful and to replace the "old" regime.

Regimes, therefore, play a central role in prevailing transition research [4]: "In the end, transitions are structural regime transformations, in which regime actors will ultimately need to change along with the process or fall out of the system" [7]. Ultimately, transitions can be seen as a regime shift resulting from a fundamental shift in power on the regime level (cf. [8]).

In order to enable transitions towards sustainability, it is critical to know the mechanisms and patterns shaping the interactions between landscape, regime and niches. A systematic knowledge of these patterns is crucial for identifying possible levers to influence transition processes, as well as concrete governance strategies and instruments [4]. With regard to global environmental change and a fair world, it will particularly become relevant to both to conceptualize the transitions between regime and landscape, and to enable and upscale system innovation.

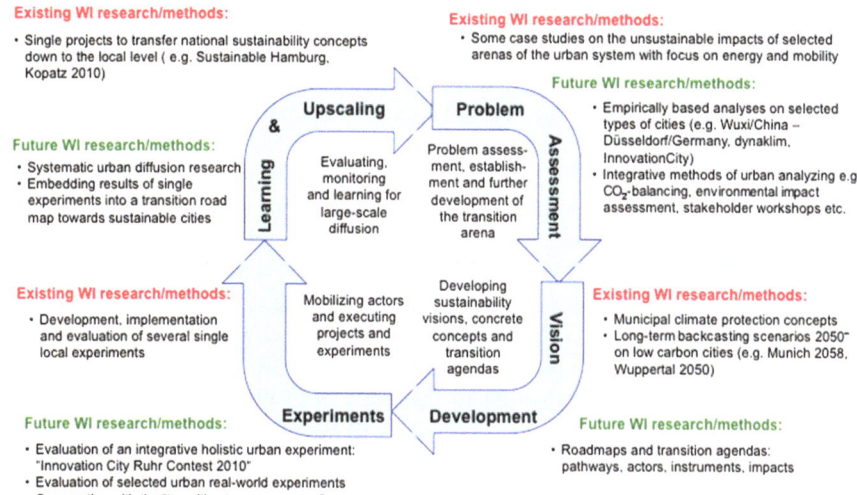

Existing WI research/methods:
- Single projects to transfer national sustainability concepts down to the local level (e.g. Sustainable Hamburg, Kopatz 2010)

Future WI research/methods:
- Systematic urban diffusion research
- Embedding results of single experiments into a transition road map towards sustainable cities

Existing WI research/methods:
- Development, implementation and evaluation of several single local experiments

Future WI research/methods:
- Evaluation of an integrative holistic urban experiment: "Innovation City Ruhr Contest 2010"
- Evaluation of selected urban real-world experiments
- Cooperation with the "transition town movement"

Existing WI research/methods:
- Some case studies on the unsustainable impacts of selected arenas of the urban system with focus on energy and mobility

Future WI research/methods:
- Empirically based analyses on selected types of cities (e.g. Wuxi/China – Düsseldorf/Germany, dynaklim, InnovationCity)
- Integrative methods of urban analyzing e.g. CO_2-balancing, environmental impact assessment, stakeholder workshops etc.

Existing WI research/methods:
- Municipal climate protection concepts
- Long-term backcasting scenarios 2050" on low carbon cities (e.g. Munich 2058, Wuppertal 2050)

Future WI research/methods:
- Roadmaps and transition agendas: pathways, actors, instruments, impacts

Upscaling & Learning — Evaluating, monitoring and learning for large-scale diffusion

Problem Assessment — Problem assessment, establishment and further development of the transition arena

Vision Development — Developing sustainability visions, concrete concepts and transition agendas

Experiments — Mobilizing actors and executing projects and experiments

Fig. 5.2 Transition concept in the field of urban transitions towards sustainable cities—existing and future research plans

As already mentioned, the Wuppertal Institute uses the transition cycle as frame (structuring element; cf. Fig. 5.2) and catalyser for its research agenda. Research activities on city and regional level have improved at the institute in the past, not only as of its scientific relevance, but as it has a strategic importance for international sustainable development and for the improvement of inter- and transdisciplinary research.

The institute has a long record of projects and experiences in analysing international, European and national urban spaces, and it currently faces the challenge of addressing new trends and developments of urban transformations in an integrated perspective.

Problem Assessment

Predominantly, current studies on urban developments focus either on unsustainable consequences of present trends for the society and the environment or on identifying local barriers to sustainable urban development. By now, only very few studies analyse urban ecological developments from an integrated perspective, and the Wuppertal Institute has identified the need for further research with a broad empirical basis and cross-sectoral approaches. There is also a need to conduct comparative case studies for different types of cities (growing cities, shrinking cities, Megacities, etc.) and to explore possible pathways towards urban sustainability. To close this knowledge gap, the Wuppertal Institute analyses different types of cities by comparative case studies, e.g. in Wuxi (China) and Düsseldorf

(Germany), as examples for "big and growing cities" (Wuppertal Institute 2010). Also, the networking and research projects "dynaklim" and "InnovationCity Ruhr" carry out multi-disciplinary research on innovative mitigation and adaptation strategies. This leads to an improved analytical understanding of conditions, obstacles, drivers and promoters of a climate-friendly, sustainable urban development.

In the future, further elements should complement the integrated analysis already done at the Institute: First, economic analyses of regional value chains provide insight into current strengths and weaknesses of regional economics in order to develop these in a more sustainable way. Second, the social dimensions of mitigation policy at the local level will be an integrative part of the analysis: In what sense will social groups be affected or benefit from mitigation policies? Third, beside mitigation, adaptation to the impacts of climate change and their synergies and conflicts to mitigation policies will be one component of institute's research on urban developments. And finally, the integration of mitigation, adaptation and resource efficiency at a local level appears as a main challenge for urban transition research, particularly in Asian countries as well as the combination of GHG mitigation strategies and air quality initiatives.

Vision Development

In Germany, there have been several efforts towards developing visions for transitions of cities towards sustainability in the last decades, as the research programs "ecological renewal of cities" (about 1985–1995) or "sustainable urban development" (about 1995–2005). Since 2005, the focus of the debate has shifted more towards "resource efficient cities" (see e.g. [9]); "climate-friendly cities" (e.g. [10, 11]); and "low carbon cities" (e.g. [12–14]). But this discussion is often about short- and mid-term "reality-grounded" concepts with inadequate solutions regarding necessary long term (structural) changes.

Therefore, research on urban transitions will continue to be based on (1) "municipal climate protection plans" and road mapping processes developing mid-term alternative scenarios and policy packages for a sustainable urban development; (2) applied research addressing the interplay between formal and informal planning instruments in local mitigation and adaptation strategies; and (3) long-term back casting-scenarios (2050 and beyond) for near CO_2-neutral cities. The Wuppertal Institute has conducted two pioneering scenario analyses on low carbon cities: "Munich 2058" [11] and "Düsseldorf 2050" (ongoing 2011–2012). This work will be continued and extended, leading to a deeper understanding of "roadmap concepts" or "transition agendas for urban development". Actor-oriented case studies on pathways of transitions for selected problems and types of cities (see e.g. [15]) should tackle questions like: Who can contribute and how? What are the steps? With what measures and instruments? And, what are the impacts?

Experiments

In the application field of urban transitions, a broad variety of experiments exists in Germany. Ranging from model, pilot and demonstration projects with a selected local or thematic focus, up to "urban real-world experiments" with complete cities (e.g. growing cities like Hamburg or some shrinking cities in Eastern Germany or in the Ruhr region). Furthermore, several sectoral experiments like car-free housing, traffic calming, 100 %-renewable energies, etc. do exist. On a larger scale, a couple of big international construction exhibitions (Emscherpark, Eastern Germany, City of Hamburg) and some sustainable cities contests have to be mentioned (e.g. Innovation City Ruhr, European Green Capital). Only in some cases, these experiments have been combined with scientific evaluation studies. Mostly, they are just a best practice collection without an adequate, empirically based analysis. In several cases of such experimental settings, the Wuppertal Institute has been or is involved, e.g. living without car [16], car free mobility in NRW [17], the YOU-move.NRW campaign [18], solar and save projects, KURS 21 educational partnerships, and others.

Currently the institute is extending its scientific work with regard to conceptualization and evaluation of real-term experiments in urban areas. Beside the "InnovationCity Ruhr" project, the framework program "Energy Transformation in the cities of the Ruhr Valley" builds a perfect scientific basis for this approach.

Shaping sustainable urban infrastructures is a complex process and experience is scarce. The project "Innovation City Ruhr" was started as a multi-dimensional and transdisciplinary real world experiment, which has a crucial role in the transition cycle. The aim is to learn more about achievable urban transition goals, socio-economic system interactions, alternative transition options and their characteristics and impacts.

The Innovation City process was launched by the "Initiativkreis Ruhr", a network of the 70 biggest companies of the RMR, as substantial contribution and accelerating moment for a dynamic climate protection path in one of the world's most industrialised region. In November 2010 the city of Bottrop located in the middle of the industrial heart of Germany won the competition and was appointed "Innovation City Ruhr". The aim of the city is to cut its GHG emissions in a representative district with about 69,000 inhabitants by 50 % within 10 years. Additionally "better living" conditions shall be achieved.

Compared to many other projects "Innovation City" is going a step further. Society and technology are inextricably intertwined and science has the role to support the transition process, to help developing experiments and to trigger learning projects. "InnovationCity" is an experimental setting in a socio-technical context where methods of Real World Experimentation [19] are employed by the city in order to trigger a complex transition process [20]. One part of this process are the so-called living labs, which involve citizens in innovation and development, thus catalysing a democratic and open innovation system. The comprehensive accompanying research programme is being organized by one partner of the Research Network. The project brings together different disciplines and experiences. It works as a transmission belt between researchers, the planning team, investors, and political decision makers. Thus, research provides first-hand information and insights being absolutely crucial for the national and international transfer of practical experience.

Additional tasks ahead are cooperation with, e.g. the "transition town movement" in order to facilitate the proliferation of local resource efficiency experiments. Visions of new models of urban development will be created, which integrate the physical and technical side of service provision and the socio-cultural dimensions of resource consumption. These concepts could powerfully shape the context for social and environmental innovation.

Learning and Upscaling

In the German urban practice and science, bottom-up learning processes to learn by comparison, discussion and benchmarking ("good practice") are common at the moment. Usually, practitioner conferences, practitioner journals and internet platforms, discuss local results and aim at disseminating such good practice. EU-programs also aim at disseminating the idea of sustainable cities, e.g. Structural Funds and Cohesion Funds, Interreg Programs as European Programs promoting sustainability oriented programs in European regions and cities. But there is hardly any scientific discussion on strategies for a systematic upscaling of success factors from existing local experiments. Questions about how to organize the learning and diffusion process in society and politics have been not answered yet.

Against this background, future research will analyse conditions for upscaling, provide systematic contributions to an upscaling and dissemination of successful local good practice, proofed on scientifically sound evaluations. Such "urban diffusion research" demands an empirical analysis of selected subjects, solutions and cities, in order to show their specific role with regard to an overall urban development strategy towards sustainability. Therefore, a comparison of case studies is needed to benchmark the current landscape of good practice. Research should employ an institutional and actor-oriented research perspective (How to come from single local experiments to a common standard? What are barriers, promoters and actors, key success-factors, policies, institutions, etc.?) and design systematic upscaling strategies to provide the dissemination of results. Additionally, the research work further supports local or regional authorities in developing and implementing sustainable strategies through strategic consultancy and applied research (e.g. by local climate protection plans and guidelines) and through analysis, e.g. conducted on the potential of formal planning instruments to implement sustainable transition strategies at local and regional level.

With this kind of "urban diffusion research" research should also continue to expand networks, coalitions and dialogues with actors from the spheres of urban development, to foster mutual learning and to disseminate good practice.

Further Steps and Working Plan

Along the transition cycle, the research partners have already significantly contributed with many projects and at different levels to the analysis and the implementation of sustainable urban transitions. However, in the future, still remains the challenge of further integration at conceptual and methodological levels regarding:

- Quantitative and qualitative methodologies, especially in problem assessment and vision-development;
- Discursively oriented approaches and technological potential analysis;
- Cross problem oriented research combining mitigation, adaptation and resource efficiency with air quality needs (particularly in dense urban areas); and
- Understanding and modelling (via Agent Based Models) the social dimensions of sustainable urban transitions.

Against that background, the research network "The Reacting Atmosphere" serves as an ideal and solid basis to follow up existing research experience and future research interests.

References

1. Fischedick, M. et al.: Mitigation potential and costs. In: Edenhofer, O., Pichs Madruga, R., Sokona, Y., Seyboth, K., Matschoss, P., Kadner, S., Zwickel, T., Eickemeier, P., Hansen, G., Schlömer, S., von Stechow, C. (eds.) IPCC Special Report on Renewable Energy Sources and Climate Change Mitigation. Cambridge University Press, Cambridge and New York (2011)
2. IPCC: Climate change 2007. Mitigation of climate change. In: Metz, B., Davidson, O.R., Bosch, P.R., Dave, R., Meyer, R.A. (eds.) Contribution of Working Group III to the 4th Assessment Report of the Intergovernmental Panel on Climate Change. Cambridge University Press, Cambridge and New York (2007)
3. JRC: Scientific and technical reports. In: von Aardenne, J., Dentener, F., van Dingenen, R., Maenhout, G., Marmer, E., Vignati, E., Russ, P., Szabo L., Raes, F. (eds.) Climate and Air Quality Impacts of Combined Climate Change and Air Pollution Policy Scenarios. Publications Office of the European Union (2010)
4. Rotmans, J., Loorbach, D.: Towards a better understanding of transitions and their governance: a systemic and reflexive approach. In: Grin, J., Rotmans, J., Schot, J. (eds.) Transitions to Sustainable Development. New Directions in the Study of Long Term Tansformative Change, pp. 105–220. Routledge, New York (2010)
5. Geels, F.W., Schot, J.: The dynamics of transitions: a socio-technical perspective. In: Grin, J., Rotmans, J., Schot, J. (eds.) Transitions to Sustainable Development. New Directions in the Study of Long Term Tansformative Change, pp. 10–11. Routledge, New York (2010)
6. Loorbach, D., Rotmans, J.: Managing transitions for sustainable development. In: Olsthoorn, X., Wieczorek, A.J. (eds.) Understanding Industrial Transformation: Views from Different Disciplines, pp. 187–206. Springer, Dordrecht (2006)
7. Loorbach, D.: Transition Management: New mode of governance for sustainable development. International Books, Utrecht (2007)
8. Rotmans, J.: The role of interdisciplinary science in the transition to a low carbon society. Presentation on the occasion of the 2nd annual meeting of the low carbon society research network (LCS RNet) (2010)

 9. Reutter, O. (ed.): Ressourceneffizienz—Der neue Reichtum der Städte. Impulse für eine zukunftsfähige Kommune. Oekom, Munich (2007)
10. Lechtenböhmer, S., et al.: Smart City—Bausteine auf dem Weg zu einer CO_2-armen Stadt. Energiewirtschaftliche Tagesfragen **59**(11), 8–13 (2009)
11. Siemens AG (eds.): Sustainable Urban Infrastructure—Munich Edition—Paths Towards a Carbon-Free Future (2009). www.wupperinst.org/uploads/tx_wiprojekt/Carbon_Free_Munic. pdf
12. Lechtenböhmer, S.: Paths to a fossil CO_2-free Munich. In: Droege, P. (ed.) 100 % Renewable. Energy Autonomy in Action, pp. 87–92. Earthscan, London (2009)
13. Droege, P., Radzi, A., Carlisle, N., Lechtenböhmer, S.: 100 % Renewable Energy and Beyond for Cities. University Hamburg and World Future Council Foundation, HafenCity, Hamburg (2010). http://worldfuturecouncil.org/fileadmin/user_upload/PDF/100__renewable_energy_for_citys-for_web.pdf. Accessed 27 Oct 2010
14. Fischedick, M. et al.: Smart City—Schritte auf dem Weg zu einer CO_2-armen Stadt. In: Servatius, H.-G., Schneidewind, U. (eds.) Smart Energy. Springer Verlag, Dordrecht (2011)
15. Reutter, O.: Klimaschutz als Herausforderung für einen zukunftsfähigen Stadtverkehr—Strategien und Potenziale zur Minderung der Kohlendioxidemissionen. In: Bracher, T., Haag, M., Holzapfel, H., Kiepe, F., Lehmbrock, M., Reutter, U. (eds.) Handbuch der kommunalen Verkehrsplanung. Wichmann Verlag, Berlin (2011)
16. Reutter, O.: Modellvorhaben "Autoarmes Wohnen am Johannesplatz in Halle an der Saale": ein Werkstattbericht. In: Gather, M. (ed.), Verkehrsentwicklung in den neuen Bundesländern: Veröffentlichungen der Beiträge der Jahrestagung des Arbeitskreises Verkehr der Deutschen Gesellschaft für Geographie (DGfG) vom 17–19 Mai 2001 im Augustinerkloster zu Erfurt, pp. 109–128. Selbstverlag des Fachgebietes Geographie der Universität Erfurt, Erfurt (2001)
17. Reutter, O., Beik, U., Boege, S.: Ruschenburg, T, Dokumentation und Wirkungsanalyse der Kampagne "Umdenken, Umsteigen—Neue Mobilität in NRW". Endbericht, Wuppertal (1998)
18. Reutter, O.: Die Kampagne "YOU-move.nrw": Öffentlichkeitsarbeit und Projekte für eine jugendgerechte und umweltfreundliche Mobilitätsgestaltung Jugendlicher. Ergebnisse eines verkehrspolitischen Versuchs. In: Dalkmann, H. (ed.): Verkehrsgenese: Entstehung von Verkehr sowie Potenziale und Grenzen einer nachhaltigen Mobilität, Verl, pp. 259–275. MetaGIS-Infosysteme, Mannheim (2004)
19. Groß, M. et al.: Realexperimente, Ökologische Gestaltungsprozesse in der Wissensgesellschaft. transcript Verlag, Bielefeld (2005)
20. Schneidewind, U.: Wie Systemübergänge nachhaltig gestaltet werden können. Ökologisches Wirtschaften **3**, 27–29 (2010)

Chapter 6
Cross Sectional Processes and Development

Brigitte Halbfas and Christine Volkmann

As can be seen from the previous chapters, the core competences of the partners in this research network can be summed up in the following way:

- Measurement of atmospheric trace gases
- Development of highly sensitive measurement devices and analytical methods
- Design and organisation of international measurement campaigns
- Regional and global modelling
- Analysis of socio-economic interdependencies and reciprocal effects in the fields of economics, academia, technology, politics and society
- Analysis and design of structures for knowledge transfer and exploitation of research results.

The unique feature of the research network is certainly the combination of such core competences in one network which even covers socio-economic aspects. Such a network has the potential of analysing the development of climate and air quality in a complete and comprehensive way, which may result in a better understanding of the involved processes and more precise forecasts. This is our only chance to take appropriate action for safeguarding the natural resources for the sake of future generations.

However, the fact that scientists from various disciplines cooperate is not a guarantee for multilayer, high-level research—for at least three reasons:

First, all scientists involved are individuals who have been socialized in different ways and have received their academic education in culturally very different disciplines. Scientists need to specialize very much in order to be successful. However, every step towards specialization and expertise in one particular field will mean a loss of the broader view on phenomena and research questions outside one's own subject area. This is not a matter of deliberate decision but rather the result of the selective perception within the disciplines and the specific acquired

B. Halbfas (✉) · C. Volkmann
Schumpeter School of Business and Economics, University of Wuppertal, Wuppertal, Germany
e-mail: Halbfas@wiwi.uni-wuppertal.de

R. Koppmann (ed.), *Atmospheric Research From Different Perspectives*,
The Reacting Atmosphere 1, DOI: 10.1007/978-3-319-06495-6_6,
© Springer International Publishing Switzerland 2014

ways of thinking and communicating. Such specialization initially limits the appropriate communication, which would be necessary for the a successful research network. This does not only apply to the network as a whole but is also true for the individual research topics. These are interdisciplinary as well, and in the course of the development of common research questions it became clear how difficult communication can be even within a single field of research. However, poor communication may jeopardize the exploitation of the full potential which is so very essential.

The second reason lies within the choice of the involved scientists. These will include experts who are presently active, acknowledged and successful in their respective fields. Research results, for example from gender research, show that we leave much academic potential unexploited due to selection and segregation or at least only exploit them insufficiently [1]. While we have no intention at all to deny the competences of the scientists that are presently responsible, we are of the opinion that organizations can by no means afford to tap potentials only incompletely. Innovations seldom occur within the mainstream of an organization, which represents the knowledge that has become a standard, but rather in marginal areas of established academic structure [2]. Accordingly, a comprehensive approach such as the research network "The Reacting Atmosphere" must not confine itself to the scientists at hand. Instead, living up to the challenges of diversity management, we should rather strive to tap the complete potential. This is especially important with regard to gender aspects and the area of young researchers.

A third point is that money is spent on research in anticipation of short-term benefits (cf. Münch's definition of "Academic capitalism", [2]). Cognitive interest often depends on the expected opportunities to exploit results through articles in renowned journals. However, most of the research network's topics are—at least initially—less attractive in this respect because they are rather unfamiliar, more comprehensive and more innovative. It is therefore more difficult to convince scientists to commit to a long-term common objective which forces them to spend much time on internal communication processes while at the same time benefits are not guaranteed. In summary, the main task of Cross Sectional Development will be to design measures for the sustained promotion of the high quality research that was postulated in the beginning. This will be the only way to generate added value which is quantitatively and qualitatively comparable to traditional academic institutions.

If the establishment of such high-class research structures is successful, it can be expected that they will result in societally relevant, if not seminal research output. Then this output will have to be communicated to the relevant social groups.

This puts dissemination as a new, if not decisive, task: If even within the research network itself communication is impeded and obstructed, it may well be even more complex and difficult to communicate the results to different external social groups. This is all the more true since in the spirit of "Reacting Atmosphere" it is also desirable to receive impulses from the target groups for the network's further work. This relationship has already been acknowledged by the research network and is to be covered by this field of research. However, additional measures should be taken in order to transfer results in such a way that they can form the basis for politico-economic decisions and thus change mankinds attitude towards the atmosphere and their responsibilities in this regard. Again it is the objective to tap all existing—in this case societal—potentials. We will elaborate on this decisive field in a later section of this article.

As an intermediate result we can state: "Cross Sectional Processes and Development" aims at providing a basis for efficient potential tapping. It will supervise the whole process continually and will ensure transfer into society.

Our work includes the following tasks:

- Internal understanding and transfer
- External transfer and communication
- Diversity management, especially gender equality and promotion of young researchers
- Accompanying research.

Our approach also features the innovative task of "Accompanying Research": We will not only deal with tapping potentials, but research and dissemination processes as such will become a subject of scientific examination.

Within these four areas we have elaborated a comprehensive working schedule, which will have to be adjusted and developed in the course of perennial research. The programme cannot be fully described here. However, due to its great importance, the area of External Transfer and Communication will be presented in greater detail in the following section.

External Transfer and Communication

External Transfer and Communication goes far beyond classical transfer and is quite differentiated. We distinguish between the following dimensions:

- Target groups of the transfer (scientific community, political decision makers, environmental associations, national and international employers' associations, selected enterprises as trend-setters and innovators)
- (Higher) educational institutions
- Regional transfer in order to create a model region
- Thematic networking with related ideas and initiatives and with innovative research areas on the fringes of Reacting Atmosphere.

Accordingly, the following packages of measures can be derived:
Dissemination of results tailored to the requirements of the different target groups. This area includes large conferences and smaller workshops for the scientific community as well as presentations to regional, national or European decision makers, information transfer to educational institutions at different levels, and informative meetings and panel discussions for media representatives.

Education of interdisciplinary transfer agents. These agents can coach scientists for talks, discussions and lectures or transfer the research results to the target groups themselves after tailoring them to the respective.

Moderation of the communication and exchange between research association scientists, enterprises, institutions, politicians and pressure groups. This can take place either via exchange networking or the agents mentioned above could take care of the systematic development of focus groups which in particular could guarantee the reflection of relevant topics, experiences and requirements into the research network.

Systematic integration of scientific results in curricula at university level. This integration should take place in several steps from the development of interdisciplinary, voluntary offers up to the systematic supervision of the integration of relevant contents into accredited curricula.

Transfer to pupils and potential young researchers to change their future attitudes and behaviour. The following activities will help to achieve this goal:

- Development of innovative curricula
- Training of teachers
- Design of attractive teaching materials, target specific films, an interactive website targeted for pupils, a climate adventure trail or a touring exhibition.

High didactic quality will be crucial here and requires the involvement of experts in education.

Development of innovative entrepreneurial ideas. Through cooperation among scientists, regional innovative start-ups and entrepreneurship experts, entrepreneurial opportunities will be identified and exploited and innovation processes in the science association will be supported and coached. The research results may also lead to new products, processes, process technologies and business ideas.

These packages of measures are to be regarded as flexible instruments which have to be modified and developed in the course of the research process. This will be supported by "Accompanying Research". The integration of findings from the internal and external communication and transfer processes will be essential for the further development and adjustment of the measures.

References

1. Pascher, U., Roski M., Halbfas, B., Jansen, K., Thiesbrummel, G., Volkmann, C.: Berufliche Selbständigkeit und Unternehmensgründungen von Chemikerinnen/Frauen in der Chemie. Eine Handreichung zu Gründungsgeschehen, Hintergründen und individuellen Gründungswegen. Hrsg. v. Verbundprojekt "Gründerinnen in der Chemie" (ExiChem), Duisburg/Wuppertal 2012
2. Münch, R.: Die Universität im Kampf um die besten Zahlen. In: Rudersdorf, M., Höpken, W., Schlegel M. (eds.) Wissen und Geist. Universitätskulturen. Symposium anlässlich des 600jährigen Jubiläums der Universität Leipzig. Leipziger Universitätsverlag, 2009

Chapter 7
Connecting the Research Network to the Wider Public

Ralf Koppmann and Peter Wiesen

In March 2011 the University of Wuppertal, the Wuppertal Institute for Climate, Environment and Energy, the Atmosphere Research Divisions of the Institute for Energy and Climate Research at the Jülich Research Centre and the Rhenish Institute for Environmental Research at the University of Cologne established the research network under the title "The Reacting Atmosphere—Understanding and Management for Future Generations". The objective of the proposed network is to understand the highly complex regulatory cycles in the atmosphere taking into account all important parameters, to identify important atmospheric processes, to examine policies with respect to their consequences and, based on this to derive recommendations on how in a changing world targeted suggestions for improvement can be realised. To achieve this, the competences of the network partners with respect to atmospheric research are combined with research know-how in the analysis of technical, political and socio-economic processes and their implementation through appropriate policy instruments and transition paths.

R. Koppmann (✉) · P. Wiesen
Faculty of Mathematics and Natural Sciences, Chemistry Department,
University of Wuppertal, Gauss Strasse 20, 42119 Wuppertal, Germany
e-mail: koppmann@uni-wuppertal.de

R. Koppmann (ed.), *Atmospheric Research From Different Perspectives,*
The Reacting Atmosphere 1, DOI: 10.1007/978-3-319-06495-6_7,
© Springer International Publishing Switzerland 2014

The socio-political approach of the research network is linked to the fact that global climate change is increasingly threatening the livelihood of mankind and the development prospects of future generations. Man-made emissions are the most important cause of climate and weather changes. While the impacts of climate change in many regions of the world will only become discernible in the medium term, however, some serious problems with air quality caused by the input of different substances into the atmosphere and specific transport processes are already clearly evident. Prompt closely coordinated action is necessary in both areas. The understanding and quantitative analysis of interactions between the two areas is one of the key societal challenges of the 21st Century, which the research network will address.

Already during its preparatory phase it became evident that connecting the research network to the wider public was one of the most important objectives for a proper dissemination and exploitation of its outcomes. As a consequence, shortly after the inauguration of the research network a workshop was organised in Brussels on September 8, 2011 to present the network and its objectives to the wider public. This event was part of the workshop series "Grand Challenges—Answer from North Rhine-Westphalia" und the auspices of the State Government of North Rhine-Westphalia.

The "The Reacting Atmosphere": Presentation in Brussels

More than 120 participants from science, industry and politics attended the presentation of the research network "The Reacting Atmosphere" held at the Representation of the State of North Rhine-Westphalia (NRW) to the European Union in Brussels. In the presence of the NRW-Science Minister Svenja Schulze and Director Soledad Blanco (Sustainable Resources Management, Industry and Air, DG ENV, EC), the co-ordinator, Prof. Dr. Ralf Koppmann (BUW), introduced the ideas behind the network.

In Brussels (*left* to *right*) Dr. Rainer Steffens (Head of North Rhine-Westphalian Delegation in Brussels), André Zuber, (Head of Office, Soledad Blanco), Prof. Dr. Michael Scheffel (UW Pro-Rector for Research), Dr. Stefan Lechtenböhmer (Wuppertal Institute for Climate, Environment and Energy), Prof. Dr. Andreas Wahner (Rhenish Institute of Environmental Research / Institute of Energy and Climate Research, Research Centre Jülich), Dr. Gabriele Erhardt (Chief Operating Officer, The Reacting Atmosphere Research Network), Prof. Dr. Martin Riese (Director, Institute of Energy and Climate Research, Research Centre Jülich), Svenja Schulze (NRW Minister of Innovation, Science, Research and Technology), Prof. Dr. Ralf Koppmann (Head of BUW's Department of Atmospheric Physics and Coordinator of Reacting Atmosphere Research Network), Soledad Blanco (Director, Office of Sustainable Resources Management, Industry and Air of the EU Environmental Directorate), Prof. Dr. Peter Wiesen (Dean of BUW's Faculty of Mathematics and Natural Sciences, Deputy Coordinator of Reacting Atmosphere Research Network), and Prof. Dr. Harald Bolt (Member of Management Board, Research Centre Jülich)

Under the name "Air Quality and Climate Change—Making things Manageable for Future Generations" Prof. Koppmann explained the concept, visions and goals of the network, which has taken on the difficult challenge to quantitatively understand atmospheric processes affecting air quality and climate change including the feedbacks resulting from interactions with socio-economic processes. The chief operating officer of the network, Dr. Gabriele Erhardt guided through the programme. The two model regions, the Rhine-Ruhr Metropolitan Area and the East Asian Megacities, on which the research network focuses, were addressed in separate talks: Prof. Martin Riese (FZJ) spoke about the urgent need of a better understanding of climate-chemistry interactions and highlighted that the Asian Monsoon plays a key role in air quality-climate interactions. Dr. Stefan Lechtenböhmer (WI) concentrated on the Rhine-Ruhr Metropolitan Area and explained how climate mitigation and air quality aspects are tackled in order to achieve sustainable urban infrastructures. Soledad Blanco underlined the Importance of climate mitigation and air quality and stressed the importance of the networks' chosen topics for Europe. In addition, Andre Zuber (DG ENV) pointed out the need for further fundamental research.

The poster and demonstration session was also very well attended: Prof. Peter Wiesen (BUW), deputy co-ordinator of the network and co-ordinator of the FP7 Infrastructure Project EUROCHAMP-2, presented this project, which consists of 14 European members and fosters networking and strives to break down the boundaries between national research institutions and open up access to the networks facilities to a wider range of researchers. Also, Prof. Andreas Wahner (Research Centre Jülich) introduced the FP7 Project PEGASOS, which brings together 15 European member states, and looks at the interactions between atmospheric chemistry and climate change.

In addition, PhD students of the University of Wuppertal performed an experiment showing the reduction of nitrogen oxides on surfaces doped with TiO_2. A full-scale proof of this process is the aim of a field campaign currently being conducted in a traffic tunnel in Brussels within PhotoPAQ, a LIFE$^+$ project.

The "Reacting Atmosphere" at the International Conference "Planet Under Pressure" 2012

Atmospheric physicist Prof. Ralf Koppmann, coordinator of the research network "The Reacting Atmosphere" and Chief Operating Officer Dr. Gabriele Erhardt presented the research network at the international conference "Planet Under Pressure" in London (London, March 26–29, 2012). "Planet Under Pressure 2012" was the largest gathering of global change scientists leading up to the United Nations Conference on Sustainable Development (Rio +20) with a total of 3,018 delegates from various scientific fields as well as representatives from industry (energy, water, food industry, financial sector, insurance companies), NGOs, development agencies and media from over 100 countries at the conference venue and over 3,500 that attended virtually via live web streaming.

The conference took place six weeks before the UN conference "Sustainable Development, Rio +20" (Rio de Janeiro, May 2012). The aim of the conference was the preparation of recommendations for policy makers for special topics for the Rio conference.

Based on the latest scientific findings, a comprehensive description of the state of knowledge about the Earth system and visions for the future development should be developed. Topics of the conference were climate research, ecosystem research, land use, biodiversity, food and water supply as well as secure energy supply.

In an invited talk Prof. Koppmann presented the research network and participated in a panel discussion on "Tackling the air pollution and climate change challenge : a science/policy dialogue".

The conference ended with the approval of the "State of the Planet Declaration", which can be download at http://www.planetunderpressure2012.net.

The Reacting Atmosphere at the "Woche der Umwelt"

At the "Week of the Environment", hosted by the German President Joachim Gauck, over 200 selected exhibitors presented their innovative and future-orientated environmental and nature conservation projects at the Schloss Bellevue park on June 5 und 6, 2012. One of them was the research network "The Reacting Atmosphere" represented by Prof. Ralf Koppmann, atmospheric physicist and coordinator of the research network, Prof. Peter Wiesen, atmospheric chemist and deputy coordinator, and chief operating officer Dr. Gabriele Erhardt. About 550 companies, organisations, institutes and initiatives applied for participation in this proficiency show. Participants were selected by an independent jury according to eligibility criteria as quality, innovation and model character and the research network "The reacting atmosphere" was selected as one of them.

Prof. Peter Wiesen (*left*) and Prof. Ralf Koppmann, presenting the Research Network at the "Woche der Umwelt" in Berlin, 2012

The Research Network presented novel, highly sensitive measurement techniques for the detection of atmospheric trace substances, which had been developed in cooperation with industry partners. A new type of measuring device for the detection of nitrogen dioxide—supported by the Deutsche Bundesstiftung Umwelt and the EU Commission—was displayed.

The Reacting Atmosphere at the Green Week in Brussels

The research network "The Reacting Atmosphere" presented its activities at Green Week 2013 in Brussels. The network had been selected from numerous applicants by the European Commission. Green Week, the largest annual conference on European environmental policy, focused on air quality in 2013—a key issue of the research network. Although major progress has been made in recent years, air quality standards in densely populated areas of the EU are still frequently exceeded. This relates in particular to fine particulate matter pollution, ground-level ozone, and nitrogen dioxide. The European Commission currently reviews its guidelines for air pollution control in order to further improve air quality in the near future and to significantly reduce the number of transgressions of air quality standards.

Dr. Ralf Kurtenbach from the University of Wuppertal's atmospheric chemistry group presenting a small photo reactor at the Green Week 2013 in Brussels

At Green Week 2013 the research network "The Reacting Atmosphere" presented a small photo reactor, which breaks down air pollutants into harmless substances by titanium dioxide.

Titanium dioxide accelerates chemical reactions if UV light is irradiated. Therefore it can be used as a catalytic converter to make air cleaner. With the involvement of BUW the applicability of this process is tested in the framework of the large European research project PhotoPAQ (Demonstration of Photocatalytic remediation Processes on Air Quality: photopaq.ircelyon.univ-lyon1.fr/).